THE FIELD GUIDE SERIES

TOM BROWN'S BESTSELLING GUIDES TO WILDERNESS SURVIVAL, WILD AND MEDICINAL PLANTS, AND OTHER TOPICS

The Tracker Tom Brown's classic true story—the most powerful and magical high-spiritual adventure since *The Teachings of Don Juan*

The Search The continuing story of *The Tracker*, exploring the ancient art of the new survival

The Vision Tom Brown's profound, personal journey into an ancient mystical experience, the Vision Quest

The Quest The acclaimed outdoorsman shows how we can save our planet

The Journey A message of hope and harmony for our earth and our spirits—Tom Brown's vision for healing our world

Grandfather The incredible true story of a remarkable Native American and his lifelong search for peace and truth in nature

Awakening Spirits Tom Brown shares the unique meditation exercises used by the students of his personal Tracker classes

The Way of the Scout A collection of stories illustrating the advanced tracking skills taught to Tom Brown by Grandfather

CASE FILES OF THE TRACKER

TRUE STORIES FROM AMERICA'S GREATEST OUTDOORSMAN

Tom Brown, Jr.

BERKLEY BOOKS, NEW YORK

A Berkley Book
Published by The Berkley Publishing Group
A division of Penguin Group (USA) Inc.
375 Hudson Street
New York, New York 10014

This book is an original publication of The Berkley Publishing Group.

PRINTING HISTORY
Berkley trade paperback edition / December 2003

Library of Congress Cataloging-in-Publication Data
Brown, Tom, 1950–
    Case files of the tracker : true stories from America's greatest outdoorsman /
Tom Brown.
        p. cm.
    ISBN 0-425-18755-1
    1. Tracking and trailing—United States—Anecdotes. 2. Brown, Tom,
1950– I. Title.
SK283.6.U6B76 2003
599.147'9—dc22   2003060888

PRINTED IN THE UNITED STATES OF AMERICA
10   9   8   7   6   5   4   3   2   1

In Memory of

MICHAEL J. HORROCKS

First Officer, United Flight # 175
March 24, 1963 to September 11, 2001
My brother-in-law, my cherished friend,
and my hero.

First and foremost, this book is dedicated to my wife and best friend, Debbie. My Vision has grown into our Vision. She is the driving force and inspiration behind the Tracker School, my books, the Tracking Search and Rescue Teams, and most of all the Vision of saving what we have left of our Earth. It was Debbie who finally convinced me to write the "Case Files of the Tracker" and create this first book of the trilogy. She is also one of my most cherished Trackers, a veteran of many tracking cases, who now stands on the threshold of being one of the best.

This book would not have happened without the full support, love, and understanding of my family. They endured while I was away tracking and teaching, yet their love and faith never faltered for this Vision is their Vision. To my sons, Coty, River, Paul, and Tommy, my daughter-in-law, Darlene, and my grandchildren, Anthony and Haley, I dedicate this book and my heart to you, for giving me the freedom to live my dream, and our dream. I also dedicate this book to the Horrocks family, who endured so much and who I love so much.

Finally, I would like to dedicate this book to the finest instructors and Trackers I have ever had the privilege of teaching: Kevin Reeve, Billy and Christie McConnell, Ruth-Ann Colby Martin, Eddie Starnater, and Joe and Raquel Lau. Truly you are all the soul of the Tracker School and I am so proud of you for carrying on Grandfather's Vision, my Vision, and now our Vision. I would also like to thank Malcolm Ringwalt and Nancy Klein for all they have done for the Tracker School, my family, and for our Vision. To all of my students who hunger for the skill and wisdom of Grandfather and strive to save the Earth, I thank you, for without you the Vision would die. A special thanks to my editor at The Berkley Publishing Group, Tom Colgan, who helped me through one of the most difficult books I have ever written, during one of the most difficult schedules I had ever known.

In sadness I would also like to thank my dad, Thomas H. Brown, who passed away before this book was completed. Without the freedom, love, and encouragement he gave to me as a child I would never have become a Tracker.

CONTENTS

CASE FILES OF THE TRACKER

THE SCIENCE AND ART
OF TRACKING

"SHE DOESN'T EVEN know that she is lost yet," Grandfather said, his statement taking me by surprise. I followed his finger to the ground, and there was a small child's track faintly depressed into the debris. I could barely see the track, and really had no idea what he was talking about. After all, I had only known Grandfather for a few months and until this point I had only tracked local wildlife in clear soil situations. At first I had no idea why he was pointing this track out for me to see, until his statement finally sank into my mind and seemed to make sense. Still I was amazed by the amount of information Grandfather was able to discern from this one track. Somehow he knew that this child was unaware that she was lost at this point in the track. I wondered how Grandfather knew that this child was lost, since to me the track appeared to have been made hours before.

His statement should not have surprised me, for Grandfather was always full of mysterious and strange statements such as this. After all, Grandfather was a paradox, an anachronism, and a bridge that led from the world everyone else lived in to the world of wilderness that I so loved. Grandfather's wisdom spanned across time and brought the ancient world into the context of the present. He was the grandfather of my best friend, Rick, and was called Stalking Wolf, but we only knew him as Grandfather. Ultimately he became my best friend for the better part of ten years, and losing him was like losing my own grandfather. His Vision had led him here to New Jersey's Pine Barrens, far from his people, to teach a young "White Coyote." Little did I know that the White Coyote was to be me.

Through all those years he taught me to survive in pure wilderness with nothing at all. He taught me levels of awareness that far transcended all human limitation, and he led me to a vast spiritual world of things "unseen and eternal." Most of all he taught me to be a Tracker and all the wisdom and philosophy that a Tracker stands for. He taught me to read the Earth like an open book and to unravel her mysteries. He opened up the world to me in a wondrous and spiritual new way. Grandfather became my doorway, my vehicle, which took me back to the Earth, and I became a Child of the Earth again, no longer an alien to my own planet. I became a Caretaker and Healer of the Earth through his teachings.

Yet, as I now look back, very few could have learned from Grandfather, for he was a Coyote Teacher. He handed me nothing; instead he demanded that I work hard for the truth and find answers by myself. Unlike modern teaching that spells out every detail, Coyote Teaching demands a passion to search for answers. If I was without that passion, Grandfather would not teach. That is why I am forever amazed when I look back on this day, this moment of time with Grandfather. In the span of an hour, and through that little lost girl's single track, I learned everything I needed to know about Tracking and being a Tracker. He gave me the wisdom and answers needed to become a Tracker, and the passion needed to follow that very difficult path.

So there we sat, side by side on an obscure path in the Pine Barrens, an eighty-three-year-old displaced Apache and a seven-year-old little boy, an unlikely pair to say the very least. "How do you know that she doesn't know she's lost yet?" I asked. He remained silent for a long time, then said, "The Earth has told me, her track has told me, and my heart has told me." Little did I know that this question and answer were the beginning of the grandest gift of wisdom I would ever learn, at one time, from Grandfather. My question and his answer unleashed a remarkable chain of events, an avalanche of knowledge and tracking, yet we would never have to move from that first track. To this very day I have never experienced tracking like that again and I have never been able to come close to doing what Grandfather had done that day, even with being a tracker and instructor for forty-five years. Grandfather tracked that little girl without moving from the first track, and through that track he unraveled the mysteries of tracking.

"The Earth told you?" I asked. Grandfather waited again before answering, closing his eyes and listening to the faint wind in the trees. I wonder if he was trying to choose his words, or if he was searching out my heart, but his answer awakened me to things I had no idea existed. He replied with his own question. "Grandson, have you ever dropped a stone into quiet waters?" "Yes," I answered. "What happens then, Grandson?" I remember thinking for a moment, a little confused and wondering what he was really asking. I sheepishly said, "There's a splash, Grandfather." "And?" he asked. I thought again, grappling in my mind for what happens next. I had thrown hundreds of stones, maybe thousands, and I sat there dumbfounded. All manner of thought raced through my mind. I almost panicked because I could not think what comes next, what more he could be asking. Then suddenly I realized, I could see it in my mind, and I blurted out, "A wave."

"That's right, Grandson," he said, "but it's not just a wave, but a ring, that moves out from the splash. And as you have seen so often, it is followed by another ring and another and another, until it finally comes to rest on the shore. Nature is like that, Grandson," he continued, "for the whole of Creation is as the waters. When something moves in the natural world it disturbs something else. A fox that is walking in the woods creates a major splash in the waters of life. Birds send out alarm calls, chipmunk dive for their holes, mice run for cover, and all is in turbulence. Yet it does not end there. Like the concentric rings of water, it moves across the land like a ripple, disturbing more and more, until it comes to rest on the shores of a Tracker's ear. That is how the Earth told me that the little girl is now lost, Grandson," he said.

I was stunned, no, shocked and amazed at his statement. He knew that the little girl was now lost because these concentric rings of nature had told him. I could only imagine the wondrous new worlds that could open up, what I could know about the forest at greater and greater distances. At that moment I also realized that this is why Grandfather had stopped so often on our journeys. It was in the stops that he would listen intently to what the concentric rings of life and the voice of the Earth were telling him. I was inspired and I desperately wanted to ask him more, but I really didn't know what to ask! Finally, out of sheer desperation, I blurted out, "But how did you know that the girl did not know she was lost when she made this track?" Yet again there was a long period of silence before he spoke again.

Finally he said, "Because, Grandson, her track told me that she does not know that she is lost at this place." I was now even more amazed and asked, "But how did the track tell you?" Grandfather answered, "The body of any human or animal is always struggling to remain upright, to stay balanced, as it moves across the Earth. To do so there is a system of checks and balances. Each movement, no matter how small, must be compensated for." "What do you mean?" I asked. He smiled. "Close your eyes and touch your nose with your hand, but pay attention to what your entire body does." It wasn't an answer at all but I immediately did what he asked and suddenly my question was somehow answered, but I didn't know how. Overjoyed, I said, "My whole body moved!" "Yes, Grandson," he said, "your body moved to adjust and compensate for your movement so that you would remain seated upright. It is the same with all things and with all movement."

He did not wait for me to ask the next question. "Tracks are not lifeless depressions in the Earth, Grandson. Tracks are more than a window back in time; they are the window to the very soul. In all tracks there is a tiny universe where all of the body's movements are recorded. It is like a little landscape, of hills and valleys, ridges and peaks, pocks and domes, thousands of them, each one indicating some small movement in the body and mind. Together my people call these little landscapes pressure releases, and there are thousands of pressure releases to be known and understood. It is in the voice of these pressure releases that the track tells me that the little girl does not know she is lost."

I was riveted to his every word. I had never thought about a track being more than an identification mark and something to follow. Tracks were just depressions in the ground, but now Grandfather was telling me that they were more, much more. I was so excited at this that all I could do was sit there and try to digest what he had just given me. Grandfather smiled at me and then said, "I also know that this girl is thirty-one pounds, she is right-handed, she has had food but she is now growing thirsty."

He continued on and on about the little girl, as if he was watching her walk by right now. But he could read deeper than just the superficial; it was as if he were reading right into her soul, just like he had said. In a way, to Grandfather, the track was also like being able to read the little girl's mind. I felt overwhelmed. Yet I had to ask more. Again I blurted out what I thought was a stupid question. "How does your heart know she is lost, Grandfather?"

"The same way I know that in a few moments she will find her parents," Grandfather responded. I was baffled by his answer, until he continued, saying, "As I have told you, there is a life force that moves through all of Creation. My people call this force the Spirit-That-Moves-In-and-Through-All-Things, and it is this life force that I listen to with my heart. It is the life force that tells me that she is moving closer to home, recognizing even now her own backyard. It is this life force, this Spirit that runs through all of us. It runs through you and me, all of Creation, the Earth, the sky, and even to the Creator. You are part of it and it is part of you. Thus, if you know how to listen to this force through your heart, you too could feel the little girl move within you as you move within her and all other things of Earth and Spirit." In the distance I could hear very faint screams of laughter, just as Grandfather finished talking. I watched a smile move across his face as he heard the laughter also.

I was amazed. He had been right! And I had heard the echo of the concentric rings now too. Grandfather did not wait for any more questions, but said, "As you put your finger in this track, pointing to the little girl's track, you are connected to her. A track is like the end of a string. At the far end a being is moving. As you pick up the track, you pick up the string, and you become a Tracker. A Tracker follows the mysteries of life, even those that lead to the unseen and eternal worlds of Spirit. A Tracker is at once all things and all things are the Tracker. Now that you have seen the mysteries unfold it has become your journey and Vision to become a Tracker, for once you have touched the wisdom there can be no turning back from the path. Welcome, Grandson, to my world." With those words on that day my life began.

MY FRANKENSTEIN

I AM LYING on the dusty rubble of this man-made hell, trying to stop the blood flowing from my back. The gunshot wound only passed under the skin, entering from the center of my back by my belt line and traveling out my left side. Even though it entered my skin just atop my spine, I don't know, nor do I care to know, how it missed crippling me. It's all a blur of surreal images that are compounded by sheer exhaustion and pain. I am at a point where I can no longer think clearly. My only concern is that I stop bleeding and regain my strength. Yet part of my immediate concern is also for the man lying not far from me. Ironically, I am not so much worried about his well-being as much as I hope that he does not get up.

The landscape around me is a dusty heap of broken-down

buildings, old pipes, and rusted hulks of nondescript machinery and debris. Not only is this a foreign land, but a foreign and surreal landscape, born more of a nightmare than of an industrial vision. This corrupt decay is slowly being reclaimed by nature's forces. Clumps of saplings and assorted brush are overtaking much of the area, grasses are growing from the cracks and crevices of buildings. Everywhere, etched in the concrete dust, are the tracks of animals, of life, and of hope. For me, this reclamation of nature over man is a welcome vision, a much-needed relief from the hideous stress I've faced for the past few days.

The trembling in my body from both fatigue and shock is slowly ebbing. Bit by bit awareness returns to my mind. The inside of my head feels as dusty and swirling as the concrete dust devils that dance down the catacombs and recesses of these old buildings. The man lying by me has not moved. Time seems to be frozen, place no longer really has any sense of reality or context. It takes what seems like a torturous eternity to reach out and pick up the .38 police special lying not far from me. The bleeding begins again as I grasp the barrel and bring it toward me. It's obvious now that I can't move and I won't be leaving this area until my bleeding completely stops. Slowly I am realizing that I am thirsty, very thirsty. Hunger is only a distant dream, for the thirst must be satisfied before any hunger returns. Dehydration, my old friend, how well I know you, I think. My mind swoons with the thought of finding any water soon in this horrid place. Even if there was any water here it would be rancid and putrefied with chemical and biological waste.

Yet it is not the water that concerns me, nor this dusty place. After all, I've faced far harsher conditions in my life. I know that I can do what needs to be done to survive, but now I am also concerned about getting the man who is lying beside me out. If it were only me there would be no problem. I have no idea how badly he's hurt, nor do I really care. To complicate the situation further, beyond regaining my senses, stopping the bleeding, and resting up to face the long journey ahead, I really don't have any idea how far it is to a known civilization. Basically I only have a vague idea where I am. Again, if I were on my own and healthy I would not care where I was. Any wilderness is home and normally I would avoid civilization. Now I have no choice. I have to get to some inhabited area as quickly as I can, not so much for my sake but for the man near me.

I know the likes of this arid rocky desert land very well. I know that it can easily kill even skilled wanderers and I both love and respect its power. Yet now the conditions are such that any mistake could cost me my life. All I can do at the moment is to gaze motionless upon the distant horizon and watch the approach of an oncoming late-day windstorm. It's little consolation that both of us are lying in a protected area, but then again, I would not dare move very far even if I wasn't. I cannot risk more bleeding, especially in my condition. Tormented thought and deep dehydration have taken a heavy toll. The mind cannot be trusted in this state. Mistakes are all too easy to make and clear thought is well beyond my grasp at this point of space and time.

I now feel myself slipping into the place somewhere between

dream and reality, nightmare and vision. The oncoming storm both fascinates and distracts me from my prison of flesh. The exhaustion finally and fully catches up with me as I slip in and out of sleep, much-needed, beautiful sleep, only to be awakened abruptly, heart pounding, with the frightening thought of letting down my guard. I struggle to stay awake, gazing at the storm and feeling the initial dry gusts of wind that blow the dusty concrete into my nose and lungs. These searing dry gusts, like opening a hot oven, further dry my blood to a thick crust, with an equal mix of blood and powdered concrete. This powdered concrete mix, now so thoroughly embedded in my wound, is an appreciated gift of styptic medicinal quality. I think how odd it is that I hadn't noticed before the coagulating effects of this cement dust gift, but at this point everything is a gift, just to be alive is a gift, no matter what struggles lay ahead.

I cautiously shift my position so that I can see both the approaching storm and the man beside me. I also firmly grasp the gun in my left hand, which feels very strange. Hell, even in my dominant right hand any firearm feels foreign, and never part of myself, even though I can shoot any weapon with the best of them. I despise firearms, for it is the bullet and not the hunter that does the killing, usually at long distances. In this world it takes no skill to kill with a firearm. In my world, Grandfather's world, an honorable hunt would be at very close range with a primitive weapon. Right now I am glad to have the gun. It would be impossible for me to use any primitive weapon now, for I just do not have the strength.

I begin to fight the demon of sleep with thoughts of weapons and hunting. A gun is a great equalizer, I think. The hunter no longer needs to be able to track or stalk to kill. Even the most blundering hunters can bag game with a firearm. Some are so proud that they can drop a deer at several hundred yards, deluding themselves and others into believing that they are great hunters. Little do they know that they are not hunters at all, just good shots. Grandfather would say that there was no honor in that kind of hunt. I can hear the words of Grandfather so clearly in my mind: "An honorable hunt, Grandson, is when the arrow strikes its mark, the fletches should just be passing the bow hand." That kind of hunt is honorable, for it demands stealth, cunning, and movement like a shadow upon the land.

It is these thoughts of the gun and hunting that now, especially in the light of all that has happened, have become so poignant and powerful to me. I am suddenly beginning to understand the hunt in a very strange and new way. A way that has made me realize that two ways of life, two worlds, of the ancient and the technical, have collided. As a result, I have been shot in the back and a man lies several feet away, unconscious. It's all beginning to make sense now. Grandfather once said, "As a hunter you also must always move, observe, and see as the hunted, for then and only then can you become both." I understand now what had been my mistake. I had remained the hunter, thus leading to this carnage. It was only in that split second, before the gun was fired, that I became both, I became "one," and saved my own life.

Still fiercely fighting sleep, I glance over at the crumpled heap of a man who had once been my student and a brother. His real name is Mike, but to all of his covert-ops friends he was known as Nails. Yes, tough as nails, or so he and everyone else thought. He is the arrogant sort of know-it-all that dares you to teach him anything, but teach him I did, and he became my creation, my Frankenstein monster. Yet I couldn't blame myself entirely, for all I did was to teach him the lethal skills of the Scout. It was he who went over the edge, he who made all of the wrong choices, and only he who chose to kill and kill again, both legally and now illegally. He chose to walk that path, a path from which few can return with any sanity, far less any humanness or compassion. He became what he lived: a warrior, a soldier of fortune, and an assassin. He was lethal, and death consumed him.

As I lay here watching the storm, I still blame myself for being part of his creation, his madness, and his deadly choices. Basically I gave him the tools to become a more efficient killer. I taught him to track, to survive with nothing, to stalk, and to move in and out of any environment without being observed. I taught him to escape, evade, counter-track, avoid trip wires, all with no backup or equipment. Most of all I taught him the Apache Wolverine fighting techniques, which make a man more animal than human. What was lost in the translation was the honor of the Scout, the warrior's creed, where the warrior is always the last to pick up the lance. He chose instead to pick up the lance first and he lived as the lance. Yes, I helped create this monster, with skills as lethal as mine. Tracking him was almost as difficult as it would be to track myself. I wonder why I didn't see the signs of trouble years before when I first met him. Possibly, if I had seen the hideous demon that lived within him, I would have taught him differently or not at all.

I begin to think back over all the events that led up to this moment, trying to determine where I went wrong with my teaching, and why I missed that demon within him. I slip into a state of mind someplace between remembering and reliving, recalling vividly the first time he and I met. As I travel in and out of pain and fatigue, exhaustion, thirst, and the power of the oncoming storm and night, I can clearly see our first meeting, hear our first conversation. Shadowy echoes of the past slowly come into focus as memory slips into the reality of reliving the events that took place so many years ago.

Our first meeting was in a crowded Quonset hut, located deep in the heart of an obscure military base, back during the dark days of the Cold War. It was an odd assortment of people gathered there; some covert military groups from the United States, some from other countries, some other individuals with affiliations, no doubt, with the CIA, and God knows how many other secretive spooks attended that fateful class. The group had been put together by two of my friends, covert military elite types I had taught many times before. I was quite young at the time, especially to be teaching these battle-hardened and sea-soned vets, but even though I was young I had made a name for myself teaching and tracking for these higher circles of covert-operations groups. I was well respected both here and abroad for my skill as a tracker and my ability to teach. The problem now was that the guys in the Quonset hut did not know me, but they did know and respect my two military friends, Chuck and Mark.

It was obvious, especially after being introduced to the class, that these men had no respect for me or my skills. Most of them were ordered to take the class and really didn't want to attend. I was probably ten years younger than the youngest of the men I was about to train and that further drove a wedge between us. As I began to address the class the men would not quiet down. They played grab-ass, talked in small groups, and some had their backs turned to me. Even as I raised my voice I could not get their attention. It was obvious to me that this would be a tough group unless I could get their attention. Finally Chuck, my old and respected military friend, stepped in and commanded the group to come to attention. Like magic the room quieted and the men turned to attention, obviously not wanting to piss off Chuck. Chuck said, "Listen to the Tracker. His skills will save your life." With that the room fell into a painful silence and I approached the podium again to speak.

As I began to address the men I could feel the disrespect oozing from them, especially from one guy in the back of the room who was snickering under his breath as I spoke. The more I spoke, the more he snickered, and I could clearly see by his actions and body language that he was full of himself. My eyes met his and he glared back at me with a look that could kill. It was obvious that he was respected by the other men, a sort of ringleader, and the men would do what he wanted. I knew that he was going to be trouble; in fact, I knew that the group would be trouble unless I could immediately gain their attention and ultimately their respect. As the commotion in the room began to increase again, I looked over at Chuck and he winked at me. Instantly I knew what he wanted me to do, what I always do when faced with situations of disrespect: I earn that respect by showing them what they don't know.

Without giving it a second thought I bellowed out to the class, saying, "I want you all to hunt me!" A stunned silence fell over the room and the men looked dumbfounded. The ringleader, Mike, asked sarcastically, "What do you mean, hunt you?" Those were the first sarcastic words I ever heard Mike say to me, and that began our first meeting. I said again, "I want you to hunt me, find me, all of you at once, and bring me back here." Another man named Shinn asked me what the rules were and I laid them down confidently. I told them that I was going out into the "Triangle," which was a five-acre thicket of woods surrounded by three paved roads. I then asked them to give me only a five-minute head start, but with one rule. I told them if they felt tapped on the back of the head with a small stick they would immediately return to the hut without saying a word. They all agreed, laughing wildly as I left the hut.

I could hear the thirty-four men charging the forest Triangle, whooping and laughing, cracking jokes about how easy this would be. I could clearly hear one guy say, "One guy against all of us, without camouflage, this is almost too easy." They ran into the forest with the same kind of charge that I expected them to make. They searched forever but couldn't find me, and one by one, sometimes two by two, they returned to the hut, their attitude now sullen. It was beginning to dawn on them that in war they would have been dead. It was also obvious to them that they were not as good as they thought they were. They looked hard but none of them ever saw me, not even after being touched or hit by the stick. I decided to leave Mike for last and torment him out of the Triangle by touching him with sticks every time he turned around, but he never saw me. He just grew more and more enraged with each passing touch.

As I thought back to that game I realized that this was the point at which the first seed of revenge was planted in Nails's mind. I knew from the start that he was the instigator and ringleader of the whole bunch, so I purposely designed my tactics so that the men right around him would be eliminated from the game, right from under his nose. My first order of business, as always, was to camouflage myself to perfectly fit the landscape using the mud, clay, dirt, and soil of the land. This camouflage is so precise and so unlike the military camouflage the men are used to seeing that it renders the wearer virtually invisible. Coupled with my ability to move in ways that are out of context with expected human movement, no one could find me, even though in most instances I was only inches from them.

I would blend with the brush, become a stump, hide in trees, or lie in the muck along the streambeds and ponds. Each time a man would pass I would gently reach out and tap him on the head or shoulder with a stick. He would whip around, trying to see me, but I would just as quickly and deftly disappear. I was using the ancient Scout methods of escape, invisibility, and camouflage that was so foreign to modern approaches, and the game, as always, was in my favor. For effect, the second-to-last man I took out was following Nails across a small log that spanned a mud hole. Little did he realize that I had become part of the log, and as soon as he stepped on me I pulled him deep into the mud. Nails turned around quickly, only to find that his friend was nowhere in sight. I could see then, at that very moment, the thirst for revenge deeply embedded in Nails's eyes.

This game I play with any cocky or arrogant warriors is something I have been doing to gain respect for many years. It was not my invention. The credit must go to Grandfather, for he would have me hunt him many times in this way. Actually the process of the game is so simple that it should be illegal. Simple for the hunted, that is. Grandfather would often say, "The hunted always has the advantage." Given the skills, cunning, movement, and invisibility taught by the Apache Scout training, no one can find me. To the hunters, the searchers, it is as if I've disappeared off the face of the earth. It is only when they blunder right by me that I touch them with the stick, and even then they rarely see me when they spin around to look. This process of elimination easily freaks out their minds and crumbles their arrogant confidence. Basically they learn what they don't know and fully understand how vulnerable they have been. It also becomes obvious to them that they are not part of the wilderness, just passing through it like aliens.

I walked back into the Quonset hut to a captive audience

sitting in stunned silence. No doubt I now had their rapt attention because I had gained their respect and more. Chuck turned to me and said flatly, "Were you tired, Tom? It took you thirty minutes to throw this bunch out of the woods. That's nowhere near your record. Oh! I know, you were just toying with someone." With that remark I saw Mike's face flush and disappear into his hands. It was obvious that he was really humbled, but also shaking with a barely controlled rage. I also knew that this rage inside him was directed right at me, yet ironically, so was his respect. He didn't give me trouble for the rest of the class, but he didn't talk to me either.

I explained to the men that this class had everything to do with what had happened in the Triangle and all of the mistakes they had made. I told them that if I'd had a knife and they'd had guns and this was a real war, they would all have been slaughtered. After all, they all admitted that they had never once even caught a glimpse of me, and when you can't see someone, you can't shoot to kill. From these opening words I moved the class through a week of counter-tracking, escape, evasion, survival, camouflage, and invisibility. I taught them to move like shadows. Slowly but surely as the week progressed they began to become more silent and cunning. I made many friends, and during the breaks I would teach some of the men how to make an arrowhead or use a bow. Some learned to make fires with hand drills and even create baskets. In the end, they all loved the class. All, that is, except Mike; I just couldn't tell what he wanted. I did admire his drive and passion for learning, however. It far surpassed normal learning and bordered on obsession.

On the final day, Mike came up to me and gave me a broad smile. He thanked me and apologized. Yet I could still feel a deeply set anger toward me, as if he wanted to test me. I didn't trust him, for there was something strange hidden in his eyes that I did not like. Finally all of the students had pulled out and I was left at the hut with Chuck. After a short planning session for the next class, I watched Chuck drive away. It was then that I felt the tug of Inner Vision, a sense of danger from above. Without turning around to look, I simply said, "Come on down here, Mike, you'll slip off and bust your ass!" When I heard his feet hit the ground I turned to him and smiled. Dumbfounded, he looked at me and said, "How did you know I was up there?" I answered by saying, "There are far greater powers of awareness that you cannot yet grasp and that will take time." Mike asked, "Will you teach me?" "When you're ready," I said. I watched him walk away into the night, dazed and confused, and a bit angry as he wondered how the hell I'd known he had been on top of the hut. I watched him disappear and then I disappeared, wondering if and when I would ever see Mike again.

Months passed without my hearing much of anything from any of the men I had trained. That is not odd, especially since these men were probably heavily involved in Vietnam, and not in the usual way. These men were covert operatives, and my security clearance, at that time, was on a need-to-know basis. At best, I would receive a cryptic note or letter, unsigned except for a nickname, and usually with a strange postmark. Once in a while I would get a strange phone message that my folks would receive, but usually these were as odd as the notes and letters. It was obvious from these fragments of correspondence that the theaters of operation were not only Vietnam, Cambodia, and Laos, but also included other hot spots throughout the world. I don't know why so many of the men kept in loose contact with me. The only thing I could figure is that they appreciated the skills I had taught them.

In a way, these notes, letters, and phone calls were like an intricate tracking case. I love mysteries no matter how the tracks arrive or where they take me. Often I would piece together reports I had read in the newspapers or heard on the radio and wonder if one of my students was involved in these hot spots I was learning about. The ones that concerned me the most were the suspicious deaths or disasters that would happen throughout the world. Many of these suspicious incidents, especially where very covert operations had been used, seemed eerily similar to the methods I used when teaching the men. Yet the most disturbing by far were the messages that I know came from Nails. These were not as cryptic as the rest, nearly always contained bragging, and were sometimes even threatening to me in a veiled sort of way. It was as if the student were testing the master.

Over the years these letters diminished in frequency, but they remained frightening. I could clearly see that Nails was crossing the line toward insanity. It seemed that he could not separate reality from nightmare, and began to live a strange existence. During the few quick, intensive training classes I later held, he tended to avoid me. When the subject of his letters or what he was doing surfaced, he withdrew into himself and gave me an evasive answer, sometimes no answer at all. He would completely ignore the question altogether, like it was never asked. It was also during these training sessions that I noticed the classic and horrid look in his eyes. His humanness seemed to be fading and was being slowly replaced by an animal or demon.

Many years after our last training session together I finally heard that Nails had been discharged. The guys who knew him didn't say much at all about him, just that he had quit years earlier and gone off on his own. They alluded to the idea that he was taking contracts on his own, a soldier of fortune, but they really couldn't be sure. They did say that he was not the same man they had once known. He had changed, become more violent, and spent vast amounts of time alone. He had withdrawn completely from everyone, even from what he once was as a man. In fact, based on the tone of his onetime friends, no one wanted to be around him anymore. Hearing this news was sad, yet it is exactly what I had expected. His notes, actions, and especially his eyes spoke of a man who was lost unto the world and himself in some horrid nightmare.

It was nearly a year after I learned that Nails was gone, figuratively and literally, that I received a strange call from someone claiming to be from a foreign land and loosely associated with the State Department. The caller only gave his first name and told me that he had worked with me twice before, both training and on tracking cases. I groped to place the voice with a face but could not come up with any tangible connection. In fact, when

I asked him a few questions about our training, he had all the wrong answers. It was obvious that he had never been one of my students. The only thing I could say for sure was that he must have been around during one of my tracking cases; he did know a little, but not much. As near as I can guess he may have been around people I'd trained and picked up what little he knew from their conversations. I was very skeptical at best and didn't really trust him or what he was telling me.

His call was short both in duration and instructions. All he said was that Nails was on the run and I was responsible for him. He said that Nails had killed several people and now they needed me to track him down. It was obvious that they wanted to keep this quiet, for he only gave me a place and time to meet a man named Frank. Before I could even think of asking any questions he hung up, which was typical for these types of tracking cases. Three things bothered me about this tracking case. First, I was not officially asked into the case; second, I did not know this man Frank, and at best I had to travel very far to get to him; and third, I would be tracking one of my own. Tracking one of my own was a very lethal endeavor and one I was not looking forward to, even if I did decide to go and track.

It was all so cryptic. It was obvious that they wanted to sweep Nails under the carpet, and that made me very angry. It's bad enough that they put all of this training and time into a person, use them up, and then abandon them when they are no longer useful or go too far, never offering any help. They, the governments of the world, throw them away when they no longer fit their purpose. Nails had clearly stepped across the line, or at least that was what I was led to believe. I searched the papers, foreign and domestic, for recent killings, but found nothing similar to what the caller had told me. Either it was someone who wanted me to do another tracking altogether, under the veil of Nails's name, or what he was telling me was true. I ruminated for days, far too long to be able to hitchhike to the site where I was supposed to meet this Frank. As usual, I waited until I was pushed to the limits of time and destination until I made a decision.

I went. It was all too odd, all too impossible that this anonymous caller would have known of Nails and hit a soft spot in my soul. As usual, when under time and destination constraints, I begged my folks for airfare. And, as usual, they wondered what tracking case I was going to get myself into. After all, to my folks the only tracking cases I took where I needed airfare were the ones that no one wanted me near—government, FBI, police, or family sanctioned. It was those unauthorized cases that usually got me into the most trouble, trouble my folks would have to bail me out of in the final analysis. Yet those same illegal and sketchy cases were exactly the kind of tracking challenges I loved so much. Those brought out the Scout in me, the warrior, and I loved them with a passion. I could not help but take the challenge of this mystery, especially where Nails was involved. Someone knew how to bait me.

My flight went well, but getting into a foreign country was a little more difficult than I had expected. I should not have been surprised at this, though, because back then I traveled light and I was far from a fashion statement. In fact, with my long hair and beard, well-worn clothes, and even more worn moccasins, it's no wonder I had so much trouble passing customs. I'm surprised I was able to get into the country at all. The trip from the small airstrip to the ancient hotel went much more quickly than I'd expected, thanks to a good-natured old man in an oxcart. The hotel was a sleazy dump and my room was nothing more than a cot, a dresser that had only one drawer and a stained mirror, and a bare lightbulb. The bathroom was communal, shared with the other derelicts that lived at the hotel. I fit right in with this wild crowd.

This man Frank did not show up at the appointed time, or even on the right day of the week. Given the ambiance of the little town I don't think they adhered to any set schedule or time. If Frank lived there he probably wouldn't have noticed that he was horribly late. I began to wonder if the whole thing were a hoax, or possibly Frank was watching me in order to assess my intentions. Possibly he was making sure that I was alone and had not alerted anyone to my whereabouts. Just as I was about to give up and head home, Frank showed up. If ever there was a man who stood out in a place, it was Frank. He was impeccably dressed in a regulation summer suit and tie, just slightly loosened around his neck. He arrived in a black Suburban that was driven by another man, who never spoke or got out of the SUV. By the looks of the dust and grit on the Suburban they had driven quite a distance and neither looked thrilled to be there.

He didn't have to introduce himself, and he knew exactly who I was. There was no doubt that he had done his homework and had several photos of me; he was that sure of himself. Come to think of it, in the business he is in, that could have been a costly mistake on his part. I wonder how he stayed alive so long. Like Frank, our introduction and conversation was short, contemptuous, and direct. He simply said, "Mr. Brown, I'm Frank, get in the car and I'll brief you." Without hesitating I ran back to the hotel room, gathered my scant belongings and checked out. I doubted that I would be coming back to the hotel anytime soon. As I approached the Suburban I noticed Frank's walk, and it was obvious he was carrying a large handgun, definitely not regulation. I also began to grow concerned that they were going to "off" me because of some forgotten sin I had committed.

We drove a long way in absolute silence. The silent driver kept glancing in the rearview mirror, both to look at me and to keep an eye out for anyone following us. It became clear that they were not supposed to be involved in this situation. The gravel road soon gave way to a rutted desert road and then no road at all. At best we were traveling on an oxcart path across barren, harsh, and very dry landscapes. I could see the tortured mesas and plateaus in the distance, and to the distance we were headed. I began to grow concerned because I had little water with me. Normally, when not on a tracking case I could easily survive and find water even on this barren land, but I would be tracking and enough time to gather water and survive would be difficult to find.

I could see a very small and broken-down shack or house in the distance, and we appeared to be headed right toward it. As we drew closer, Frank finally spoke, but again with a conservation of words and a certain evasiveness. He told me that this was the last known place Nails had been. Nails had killed a young mother and child, holing up in this shack for several days. On the last day there he killed the returning farmer and fled into the mountains. The bodies had been ritualistically mutilated and the blood was used to write cryptic mottoes and drawings on the walls. Because the bodies were not found for a week, and given the conditions, the authorities could not be sure if there was cannibalism involved or if animals had eaten part of the bodies.

Frank and I stepped out of the Suburban and into a hot cloud of dust. There is a place inside of myself that I call the Tracker Point of View. It is where everything is enhanced. Sound is magnified, tracks become windows not only to the past but to the soul, and all senses are at the highest level of alert. Such was the level when I passed the front gate and immediately noticed Nails's footprint. Footprints to me are like portraits to most people. A footprint is as identifying as a fingerprint, but unlike a dead fingerprint a footprint reveals everything about its maker. There was no doubt in my mind that Nails had been there, and judging from the deterioration of the track he had been there nine days ago. I knew that I had my work cut out for me because Nails had a long head start.

Frank accompanied me into the house. Here my senses screamed, for I could still see the carnage splattered upon the walls, the writing, and the stench of rotted death still hung heavy in the air. The air was thick with the buzz of flies searching for small bits of carrion or alighting on the bloody walls. Here and there were written military mottoes intermingled with insane scrawlings, lacking in context or any sensible definition. Then, out of all the writings, one caught my eye and I was dazed, no, shocked. Written in charcoal was the old Apache Scout teaching, "The hunted always has the advantage," and I knew that Nails was using this to taunt me into finding him, hunting him. Frank looked at me and said, "You see why we called you? He is one of yours and now you have to find him and bring him to me, and only me."

This statement and request really pissed me off. They assumed that I created this demon and that I had to clean it up for them. As if I were responsible for creating this mess and they were without sin. That was so much crap, and I told him so. Frank was clearly upset at my anger. He said, "We didn't really want to call you in, but we lost him and his trail several days ago. We called off the search because everyone, except some of us, thinks he is dead." With that Frank pulled a gun and a flare from his back and handed them to me. He told me that Nails was definitely armed and I would definitely need a gun. I declined the offer because I never carry a weapon. A weapon to me becomes a liability. He asked me to take the flare and to torch it off if I found Nails. He'd be watching for it and come take us out.

I looked at the flare incredulously and Frank simply said, "We have our ways of locating you and that flare." It sounded like a load of crap to me and I wondered again if I was being set up. Upon returning to the SUV Frank gave me a large jug of water, and without as much as a good-bye got in the truck and took off. I watched them disappear in a cloud of dust, seeming glad to be away from this hell. On the other hand, I was right in the middle of this world and about to track someone who would be a challenge, especially if he remembered anything I'd taught him. As dusk fast approached I made a camp well outside the home site. There was no way I wanted to stay in that cruel shack surrounded by all of the insanity written on its walls. No, I felt safer with the snakes, scorpions, and spiders, and I really hate any kind of building anyway. At least being outside I could feel and hear the concentric rings of anyone approaching, which was not so within the shack.

I headed out before first light, following no track in particular but heading outside of the search area. I knew that if I were to find any clear tracks belonging to Nails they would have to be well outside the areas that were heavily searched. I instinctively knew that I was headed in the right general direction, because that would be the choice Nails had made. From what I taught him he would have gone in the direction that the searchers would least expect. It was also one of only two routes where he could throw off tracking dogs and stay out of sight of searchers on foot, horseback, and from the air. His route into the rugged landscape would also afford better survival conditions for a man on the run. All the while my senses were at high alert. I watched and listened with such intensity that by noon my head spun with overload. I had to be careful that he was not hunting me, not watching me, not circling in for the kill.

As I pushed far outside the search area a horrible and haunting thought occurred to me. I wondered if Nails were not playing some sort of deadly chess game. He would know the way I would think and possibly do the opposite. He was not a stupid man; a stupid man would not have lived this long in his profession. No, Nails was far from stupid and probably one of the best warriors I have ever known. If he followed what I taught him he would lay down false trails, backtrack, or make moves in counter-tracking that would eat up a lot of a tracker's time, time that I could not afford, at least not in these conditions or under these circumstances. Any mistake could prove deadly, so I could chance nothing. I had to think like him and as a tracker I had to become him. Otherwise I would lose this contest, this ancient game of life and death, hunter and hunted.

By nightfall I was well outside of the search area but had not found any of Nails's tracks. I had no choice but to bed down in more of a hide than any type of shelter. I had to sleep so that I could not easily be found, and be well forewarned of any approaching danger by the disrupted sounds of the night. I might just as well have stayed awake the entire night such was my tormented sleep. Every odd sound, or lack of sound, brought me to rapt attention. My heart would race with adrenaline as I searched the night, trying to figure out what had awoken me. In that place between sleep and reality I felt something's presence, something watching me, more through Inner Vision, a sense of knowing, than from any outside clues. The whole night made me uneasy and fatigue was a place I didn't want to be. Fatigue could easily compound problems in my thinking and movements, causing me to become sloppy or let down my guard.

Long before first light I began to push on, playing the ever-intensifying chess game with my thoughts and choices. I did not find Nails's first track until well after sunrise, and it was a tremendous relief. I had been right to follow my Inner Vision and my instincts. It was obvious that Nails was not yet playing any game, but sticking to the same instincts that had driven me. Now that I had a clear trail, I had to be very careful of what lay ahead. I did not want to be lulled into a false sense of security, thinking that Nails would always be following my teaching. I could not be certain if and when he would change tactics and begin the counter-game. I also could not allow my concentric rings to give away my position. If Nails was within that Tracker Point of View, I would be difficult to track and would approach him without him knowing many miles, and even days, in advance.

I don't know why it caught my eye. It was a combination of a stray sign, a glint, and the feeling that something was out of place. I was on his trail, but the trail refused to give away its secret. It was only after seeing that which was out of context with the land that I realized Nails had circled around far ahead and set a trip-wire trap. It was a hook type and very hard to spot at best. Upon closer inspection I realized the trap would not have harmed me. It would have set off a small landslide and caused tremendous concentric rings, a vast ripple through the desert that would have alerted Nails to my approach. Upon finding the trap a deeper struggle began to emerge inside of me, a more intense set of questions and rules for this chess game. Did Nails set the trap for anyone, or just for me? I began to believe that it was set for me; such was the perfect way it was camouflaged. He would not have needed that kind of set for just any searcher. This was definitely designed with the hunter, the Tracker, in mind.

At this point my pace had been cut in half. I could not decide whether to abandon his tracks or take a more instinctual route to where I felt he might be. Taking the instinctual route would be far safer, but I could not be sure that I was headed toward him. I needed hard evidence, a track or concentric rings of nature, to make sure I was still on his trail. If I did parallel his tracks I would have to cut back in occasionally to verify that I was going in the right direction, but this would waste time. The problem was, if I stayed on his trail I would have to set my pace to a crawl, just in case he had set more trip wires or traps. Every decision was a difficult one, every move had to be thought out, and every step carefully taken. I also could not assume that all he had was a handgun. That is what Frank had told me. They assumed that because the family had been killed by a .38, that's all he had. I assume nothing. After all, he could have a high-powered rifle with him that would be capable of dropping me at a great distance.

One of the most difficult tracking situations is when I am tracking a man with sniper capabilities. As a tracker your attention is divided between the close-focus world of the track and everything between that track and eternity. This sniper capability, augmented by potential traps and Nails's ability as a Scout, upped the intensity and stress factor to its highest and most exhausting limits. It was bad enough to face this tortured landscape even without having to track, but given all of the other variables, the mind, emotion, and spirit, as well as the body, become horribly beaten by this level of stress. With this sniper capability in mind, I decided to stay far outside the trail and cut back in for confirmation only when necessary. This decision would help me avoid the potential of sniper fire and avoid any traps that he might have laid down.

This decision proved to be the best, for my pace quickened, enabling me to move as fast as or faster than Nails. For a tracker it is important to move faster than what is being tracked, otherwise I would never have a chance of catching up. It became obvious throughout the day that I had made the right decision. Upon cutting back into his track I found three more traps, with no doubt now that they were designed for me. Two of the traps were designed as early-warning systems but the third trap was potentially lethal. It was obvious that Nails wanted me dead, or at least, given the potentially lethal qualities of the trap, that he didn't care. Yet even though I was now moving as fast as or faster than he, I was still far behind. I had to formulate another plan of action. I knew that I had to make another radical decision that would take me deeper into the mind of Nails, a place I didn't really want to go.

By the sixth day I emerged from the rugged mountainous landscape and gazed down upon a huge valley. I could clearly see that it was drier than these tortured rocky peaks and nearly devoid of water. Up to this point finding water in seeps and dried streams had not been a problem, but I knew that it would take me at least two more days to cross the valley. Water would now be impossible to find. Looking across the valley to the distant hills and mountains a horrible realization came over me. Nails could be watching the valley and his vantage point would provide him with a clear shot of me. This valley forced me to make that radical decision I dreaded: I had to abandon my method and fully become Nails. I would have to use all of my skill to cross the valley and find him, now more by instinct than ever before.

I could feel him out there, hiding like a lethal hunted animal. I could enter his distorted mind and thoughts to a place so dark it made me sick. With the setting of the torturous sun I began to sense where he had gone and where he might be hiding. It was part guess, part instinct, and part knowing the forces that were driving his mind, and I somehow knew what decisions he made. At full dark I left my hide and began to cross the valley, far outside his route of travel and trail. My movement led me far away from where I last encountered his track, and now there was no way I would dare return to check his trail. I was relying totally on thinking like him, acting like him, and becoming him. The place inside of me was as dark and tormented as the valley I crossed. The darkness, however, also became my friend, for I was able to cover vast distances quickly. By first light I had crossed the valley, cutting a full day off my tracking with far less stress.

I slept as best I could for much of the day, holed up in a little dry cave that was blessedly cool and well away from the sun. My sleep was tormented and sporadic; I awoke frequently from hideous nightmares, feeling more hunted than hunter. In the deepest recesses of my mind I had become Nails, for he no doubt would be having the same kind of tormented sleep. I wondered if he even knew that I was on his trail, or for that matter, if he knew that I knew about his killings. After all, I had not made any mistakes that I knew of, yet a nagging feeling from deep inside me intensified this life-or-death chess game. I wondered if he was playing the same game by entering my mind and thinking like me. I also wondered if he sensed my intentions, my decisions, and anticipated my radical approach to this nightmare. I was fairly confident that I had made the right decision because of its radical nature, but that feeling still persisted that he might know of my out-of-context approach, and lay in wait for me far up in the mountains.

I moved again at full dark, slower now, thinking like him, reacting to every nuance of the land. My line of travel took me toward the distant craggy peaks, a mesalike plateau that lay far ahead accented by intense starlight. That place would have been a good place to hide for an Apache Scout, for it was out of context, easy to defend, and a place no one would expect someone to be hiding. Nails would probably be there, or at least I hoped he'd be there, for it would not be an obvious choice for a tracker, and thus he would think it safe. Still the nagging feeling that he knew of my approach persisted, but this was an approach so radical I wondered if he would even consider it at all. After all, I had never taught it to him, for it relied far too much on Inner Vision and a deeper spiritual understanding, and I never believed he had those qualities outside of basic survival instinct. In my soul I could feel this battle between us becoming one of instinct against spirit, a brutal battle at best.

Before first light I had reached the mesa area and found another little cave to hole up in for the day. I was not yet ready to move during the day, for I had to be sure through the hard evidence of tracks that I was truly on his trail. After checking the cave for scorpions and snakes, which was especially difficult in the dark, I entered the cave to sleep. I had a very uneasy feeling when I laid my head down, for I somehow could feel his presence close by. I dismissed this feeling as a mirage born of extreme thirst, physical and mental exhaustion, and intense stress. I wandered in and out of sleep as the sun rose and began to pound the land once again. Sometime during these early-morning hours I opened my eyes and was shocked awake by a sight I could not believe. Inches from my head were his tracks. I sat up abruptly, shaking with terror, for the track was only a few hours old. He had slept there only an hour before I arrived and I have no idea why I didn't pick up his concentric rings.

Somehow I had been driven to this place by a combination of knowledge, instinct, and spirit, or possibly it was just dumb luck, but I just didn't care. There would be plenty of time to figure it all out after this was over. I did not have time for the luxury of understanding it all right now. Nightmarish thoughts began to fill me. Certainly I had missed the concentric rings of his position, but had he missed my approach? Did he abandon this cave because he knew I was coming? If he knew I was coming, then I was trapped inside this cave. I searched the ground desperately for answers. In moments I found his exiting tracks and studied them. Any mistake with my analysis of these tracks would surely lead to my death. There was not a doubt in my mind that he might be waiting right outside. I hoped that because he had been in the cave he might not have heard my approach, but I had to be sure, dead sure.

The tracks revealed nothing outside of his normal intense caution, at least at the point where he exited the cave. He could have been partway around the mesa and still heard or felt me approach, I just didn't know. I had no choice but to follow his tracks, one by one, reading every nuance, every action and reaction, until I could be dead certain that he did not know I was there. Following these tracks would put me well inside the danger zone, both for sniper fire and any traps that he might have set, for now I was back on the expected route of approach. Each track became an intense agony of analysis, but my confidence grew as the tracks told me he was looking back toward the direction that I might be taking across the valley and not the radical new direction I had taken. I could not allow myself to be lured into a false sense of security, however, because the very next track could show signs that he knew I was there.

I had not followed him more than a mile when I encountered another lethal trap. It was a carefully camouflaged trip that had been set, one that would trigger a violent landslide of boulders above me. If I had missed it I would have been wiped off the face of the mesa and crushed. I had to make the decision again to abandon the trail, for to remain there would only put me in more danger and slow my pace again. I agonized over this decision because I wanted to find out for sure if he knew I was on his trail, or if he only vaguely expected that I might be. Yes, to abandon the trail would make it safer and easier to travel again, but I would not know if he really knew or not. That tormented me because I wanted to know, I needed to know, and maybe I would never know. I just hate not knowing, even in the best of circumstances. I could only take comfort in the fact that the decision was made for me by the presence of the trap.

It was a tormented journey, compounded by the hot sun, little sleep, intense thirst, and second-guessing his route of travel. The more I thought like he did, the more tortured I became, the darker my thoughts. At least I now had one thing in my favor for sure—it was obvious that up until this point he had not detected my route of approach and had relied far too heavily on his traps to give me away. Though this gave me some relief, it was obvious from his choice of travel that he was good, really good, but not as good as I expected. As I look back now, from this foggy agony of the gunshot wound, I realize that this had been my first deadly mistake. It gave me far too much confidence, which, subsequently, softened my judgment. Yet as I think back further, my initial mistake was missing his concentric rings as I approached the mesa cave in the first place. At the time of following him, that was more of a concern for me than underestimating his ability and overestimating mine.

I gaze out upon the darkness of this concrete jungle and think back again in search of other mistakes I made, both before and after the mesa cave incident. The confidence had overtaken me and clouded my judgment as I began to move toward where

I instinctively felt he would be hiding, or possibly waiting for me. I had been correct so far in this chess game, but I also knew that both of us were approaching a point of check, or possibly a deadly checkmate. Instinct for a tracker goes well beyond a hunch; instinct is driven by the forces of becoming what you are tracking and the spiritual reality of Inner Vision. It was this Inner Vision that was guiding me now, as it had before, and not the tracks. Knowing the tracks of Nails so well now and entering his movements and mind, I knew that this was one skill that he lacked. He did not know Inner Vision or the spiritual world, he was never interested in learning it or believing in it, and that could be a costly mistake.

So many times I tried to teach the men I trained the skills of the spiritual world. Some would embrace it fully and learn its wonders, others approached it at first with skepticism but eventually came around, and a few others blew it off altogether. Nails would never entertain even the thoughts of a spiritual reality, or even the more accessible world of the Spirit-That-Moves-In-and-Through-All-Things. He would never try an exercise and I knew that he shunned it and denied it from a place of fear. He knew that if there was such a place and it was real, then he could no longer justify his actions as a soldier of fortune or an assassin. His tracks, thus far, had not revealed any indication that he picked up any spiritual sensitivity other than to react to the concentric rings of the landscape. Yet I still could not make that assumption confidently. I knew from experience that when one's life is in danger, Inner Vision hammers through. In Nails's mind, his life was in danger.

I followed my Inner Vision—driven instinct far outside his last track and headed to a small ridge. From a Scout's point of view the approach would be perfect. Not only was it in the open and in a very vulnerable position from the point of being seen, but it would be the unexpected approach. In the mind of Nails he would not believe that anyone would be so stupid as to approach from the wide open. Yet I had taught him that the Scout always does the unexpected, the impossible, but this route was too impossibly exposed. Before approaching the ridge I further camouflaged my body with the dust and grit of the land. Even though I had done the same camouflage ritual at the start of the track, I wanted to make damn sure that it was a perfect blend with the rocks and sands of the earth. Becoming invisible would be critical for this open and potentially lethal approach. Finally, as the shadows grew longer, I headed toward the distant ridge.

These lengthening shadows were a blessing, for I reached the ridge far sooner than I had expected. I moved within the context of the land, slipping from shadow to shadow, boulder to boulder, being careful not to rush or throw any concentric rings. Finally as the light of day was waning I reached the top of the ridge and stared down in a state of shock and astonishment. Far below me, nearly obscured by the deep shadow of another distant ridge and the twilight, lay the ruins of some sort of town or factory. I could not make out exactly what it had been. From my vantage point it appeared like some sort of bombed-out ruin, abandoned and forgotten by man and eroded by time. It stood out in stark contrast to the little valley it was nestled in, another monument to man's cancerous vision of dominating nature.

I lay there for a long time trying to see into the ever-darkening landscape far below. My senses screamed for relief as I searched for any evidence, any concentric rings, that Nails might have taken refuge there. It would have been a perfect place for him to hide. It afforded protection, survival, and a clear view of any approaching searchers. What was perfect for him was my worst nightmare. My element is the raw wilderness, and in any city or town my tracking ability is greatly diminished. Even though this was the decaying remains of structures, it was still not pure wilderness.

I faced a collision of two worlds, the epic struggle between man and nature, between Nails and me. Below me lay a balance of both worlds, and if Nails was part of it we would be on equal ground. It was then that I caught the slight hint of concentric rings thrown by a mouse. It was so small I might not have noticed it if I weren't watching. It was unmistakable—Nails was there, hidden and waiting.

I did not dare sleep. I had to use the full cover of night to get into the small town. The slope was so exposed that there would be no way I could approach during the day. My nerves were at their limits already and I had to go now. As I worked my way down the slope I took solace in an old saying: "On the plains of hesitation lay the blackened bones of countless millions, who at the dawn of victory lay down to rest, and resting, died." I had made the right decision. I could not rest or hesitate. Conditions were in my favor, the night was in my favor, and Nails would not be expecting me so soon. But my fatigue and deep thirst were a liability and could obscure judgment and instinct. I did not know how rested Nails might be, for the hunted always has the advantage.

The arduous journey down the slope was far more difficult than I had imagined or anticipated. It became a battle between moving carefully and reaching the refuge of the slope's bottom before first light. If I did not reach the boulder-strewn base by first light Nails would see me. There was no way I could spend the day on the slope, for there was no place to hide from Nails or the sun. I also had to take great care not to leave any tracks. Even a moderately trained tracker can see tracks leading down a gravel slope from great distances. Every step had to be carefully chosen and placed, more by feel than by sight. As the race between the two extremes of careful movement and approaching light intensified I began to move more by Inner Vision–driven instinct than by mind and body. I was so fatigued that I had to fully trust this instinct. To do otherwise would mean failure and possibly death.

I reached the rock-strewn base of the slope just in time. As I slipped into the refuge of this boulder field the rising sun hit the slope I had just descended. Some of the boulders were as large as small cars, offering me many hiding spots. But my relief was quickly eroded as I looked back up the slope. A few of my tracks were obvious. Certainly I could see them, but could Nails? It's sometimes difficult to separate my ability from that of the people I've trained. I always assume that people can see what I see, observe what I observe, and in a way it's a blessing to overestimate. To me a mouse track on a solid rock is easy to see, yet most trained trackers do not see them at all. I prayed that Nails would not see the disturbances of my tracks on the slope; after all, he had only been tracking for less than a decade. I was able to relax a little more when I saw a collared peccary disappear along the ridge. Surely anyone would mistake my few track disturbances as those of the peccary, I thought.

I decided finally to rest and try to catch a quick nap if I could. There was nothing much I could do about my intense thirst and hunger. Though this boulder field afforded plenty of small cool caves to get me out of the sun, there were no plants or even dried seeps where I might dig for water. I had not had water in a long time and this deep dehydration would affect my body and mind in a negative way. As my body slowly consumed itself, mirages would be born and paranoia would become a reality. Now my struggle moved inward. I had to push through the mirages and nightmares of my mind and grope for reality. More than anything, I needed to call forth the animal within and become more action and reaction, more instinct than intellect. It would only be by letting go of myself fully that I could ever hope to stay alive. I could no longer think like a man but had to react with the instincts of both man and animal.

As I lay in my small protected cave I began to think of the struggles that were ahead. My concern shifted from finding Nails to determining what to do once I found him. After all, he was armed and I wasn't, he had time to rest along the way, and he was the cursed hunted and had the advantage. This was a missed vision for me. It should have been then that I realized that I too must become the hunted as well as the hunter. After all, in a way I too was being hunted. The hunter lay in wait for me, like a deer hunter sitting in a tree stand waiting for the deer to pass by. I was deer, the hunted, but I didn't realize it at this moment. Clearly this was a horribly lethal mistake. All I could think about was how I would put Nails at the disadvantage and get the gun, or guns, away from him. Then it would be one on one. He was a military-trained combatant and very lethal and I was Scout trained, lethal too in the brutal aggression of Wolverine fighting. Normally I would have the advantage, but now I was beyond exhaustion and he was rested. It would be an equal battle if it came to that.

I began to think desperately through all my options, all the skills I had learned from Grandfather along with those I had gleaned from life. I did not want, nor could I chance, a one-on-one fight with Nails. I would have to get his guns and subdue him somehow. Possibly I could even trap him. Whatever I decided to do it had to be done with all of the skill I possessed, all of my memory, training, and instinct. I had to have a plan before I reached the ruins, before the final moves of the game were to be played, but all of this depended on whether Nails was in fact still there. Possibly he had planned the same thing for me. He may have intentionally just passed through these ruins to lure me there, to set traps of his own, and to slow my progress. If he was still inside the ruins, which I now expected from the faint concentric rings, he would have set traps for me. These traps, whether he was there or not, would be highly camouflaged, precisely set, and deadly.

I began to second-guess whether I should go into the ruins at all, at least not until I was sure whether he was there or not. The prudent thing to do would be to circle the ruins and see if he had gone on, or in fact if he had been there at all. If I did decide to cut track and found that he had gone on I could follow him, avoiding this tortured place altogether. I would also know if he was really inside or not. This way I would not have to put up with any tracks or decrepit buildings. I would gain vast amounts of time. On the other hand, if I could confirm that he was there I knew I would have to move with the utmost caution. But my instinct overruled these conflicting thoughts. I knew deep inside that to hesitate would only give him the edge he needed, further depleting my strength and intensifying my thirst. I would be more prone to mistakes, for it would take me two full days to circle the ruins without being seen. I chose to follow my instincts and training, stop second-guessing myself, and go in.

I had only a few moments of fitful sleep but was able to rest up. I decided to move in at high sun, doing the unexpected again, when most people would have remained hidden and moved only under the cover of night. But I was not like everyone else. The ancient Scouts would have done the same, so would Grandfather, or at least I thought. As I began to move I wondered if Nails would expect me to come when the sun was high. It was the same as I had taught him to do; under these circumstances it would be the least expected move. With the training Nails had from me he would expect me to move when the shadows lengthened, not at this time. At best, any move at any time was a gamble. The game played out in this arena was the grand struggle of flesh and spirit, hidden in this forgotten valley, with only the eyes of mice, scorpions, and peccary to bare witness.

I moved to a place near a cluster of boulders, taking care not to leave any tracks. I needed to look at the ruins and determine what they were and their layout. I had to know, or sense, where Nails would be holed up, if he was there at all. I had to find more tracks, more evidence, and more concentric rings, without giving away my presence. As I moved to the boulders, choosing my concealment carefully, I gazed toward the tormented structures that lay before me. This was not a town but some sort of factory. There was a large semi-demolished building at its center, and strewn all around this larger structure, seemingly at random, were many other fallen-down secondary buildings, many just rubble. Old semi-paved roadways, like small city streets, crisscrossed the rubble, disappearing and reappearing. Twisted, rusted, and decaying metal, old machinery, spent autos, and various bits of debris intermingled with the fallen mass of cement and rocks. It was a factory once, that was obvious, but what kind I did not know.

I looked long and hard at the rubble, pressing my senses to the limits of my endurance. As I was about to move from my vantage point I caught a movement. It was quick, nearly invisible, but it did send off faint concentric rings. I knew that Nails was inside. Only he, or someone I had trained, could have left such a small disturbance. My senses now rocketed to a higher level than before—every nuance was scrutinized, vision intensified, colors enhanced, and even the sound of mice and scorpion feet were easily heard. I smelled and tasted the air to see if I could sense the lingering scent of him. Even my sense of touch made every nerve come alive as if electrified. I watched every flaw and disturbance in the landscape, every bent blade of brown grass, every pebble that lay out of context within this rubble that itself lay out of context with the land. I knew where Nails had gone in, and now I was damn sure that he was still there. My heart raced, for the confrontation was close, and now every movement, every second, had to be calculated.

I knew now why he chose this place, but it would not have been my choice. This was what was familiar to him; the rusted metal and broken-down structures afforded him a grand survival potential for his skill level. It was easy to defend, and no doubt easy to find water. There had to be an abandoned well somewhere nearby. I would have chosen to go to the craggy rocks and caves far up the ridgeline. There I would have been at home, just as the ancient Scouts had been, and just as Grandfather had taught me to be so many years ago. Rugged and pure wilderness to me, to Grandfather, and to the ancient Scouts is always a refuge, a place called home. It is wilderness that keeps us safe and free, where we can roam outside the worlds of man. Nails had chosen something close to his world, his home, but my home was encroaching upon it; right now there was a balance between the two forces, but soon it would be won back by the forces of Creation. It became for me the razor's edge upon which I must balance, being able to see my world and his world at the same time.

I began to move like a gentle breeze that disturbs not a blade of grass. I clung close to the larger boulders, becoming their shadow, taking care not to unsettle even an ant. It was not long before I reached the outermost fringe of the rubble. I had chosen the most open route I could find. I knew that Nails would be looking in the more obvious places for any approach. I began to study the landscape and the ground at greater distances, trying to find his tracks long before I would come upon them. This would give me an early warning of his movement up until now and alert me to any traps he might have set. I learned long ago that a tracker has to see tracks from great distances, especially when tracking a fugitive. Early warning, even when tracking a lost person or animal, is critical to awareness and safety. I also had to be extremely careful to leave no tracks, at least no tracks that he could see, and I had to remain invisible, just in case he happened by me.

Finally I found his track not far from where I lay and a great sense of relief washed over me. He had arrived there only a few hours before me and that would not have given him enough time to set many traps, if any. He was probably still carefully exploring his new home and, given the nuances of his tracks, he did not seem very concerned about anyone following him. But his tracks did indicate that he was concerned about the abandoned buildings, as he should have been, as he had been trained to be. Upon his initial approach he would not have known if anyone occupied any of the buildings. His attention would have been diverted away from what lay behind him and turned toward what lay ahead of him. It should have been both. Grandfather had taught me, as I had taught Nails, that attention should be all around, not just ahead or behind, but all around. Nails had broken a simple rule. Now I was breaking the same rule, rudely awakened by the fact that a rattlesnake had just rattled at me not far from my side.

I lay still, frozen in time, slipping into the Sacred Silence of the soul, in a place of splendid nonexistence that creates a state of invisibility to all that passes by. The snake slowly slipped by just a few inches from my head and into the cover of a crevice. As he disappeared out of sight I began to go into a state of panic. The rattle, I thought, even though short and quiet, did it throw the kind of concentric rings that Nails would pick up? I couldn't chance it, I had to get out of this position, and the only escape lay within the confines of the rubble ahead of me. Going there would take me into Nails's home, but I had no choice. If he did hear the snake he would come to investigate. Not only did I have to escape and find cover but I had to be more careful than ever before not to leave any tracks he could find. As I moved I fought the adrenaline rush, for to give in to it would cause even more disturbances, more concentric rings, and more tracks. It is so difficult to fight against this primal fight-or-flight instinct that drives us all. The ancient Scouts were trained to move out of that state and do the unexpected, the impossible.

I found refuge in the rubble almost immediately with little effort. No doubt the Creator and the spirits of the Scouts were with me because it was the only place I could have gone. Any other choice would have put me in the open. I held my position for a long time, waiting, watching, and intently scrutinizing every nuance of this strange landscape. I had to be sure that Nails had not heard the rattle. The sheer intensity of the waiting and watching began to take its toll on my senses. I could feel the thirst, the aches, the pounding fatigue, and I had to struggle not to retire within myself, for that would be the place of mirage, a coffin of flesh, cut off from what is real and dangerous. I fought to stay alert and ready, like a hunted coyote, just in case I missed something while making my initial escape. I had to pay attention to everything around me too, for I could not chance disturbing another animal, no matter how small and insignificant its concentric rings. I had to move outside the realm of even the smallest of creatures, a task that is not easy to do even at the best of times.

Moments had turned into what seemed like hours but the sun had not moved that far across the sky. I pressed my senses far outside their limits, not only focusing on any movement within the ruins but well beyond. I could not chance the assumption that Nails was still inside the rubble, for he might be foraging for survival in the surrounding mountains now. It would be a horrible mistake to make my move if he was in fact at a higher elevation outside the area. From a higher vantage point he could clearly see anyone entering the factory area, no matter how well camouflaged. Any concentric rings thrown from the rubble would intensify as they traveled up the face of the mountains. I first focused on all of the areas that he might be moving or hiding in and then refocused my attention on all of the areas he would never use. It was there, in a totally unexpected area, that I picked up a very faint sign of his presence. It was nothing more than a small pile of rocks and broken cement near a crevice of a fallen building, but it marked his first camp.

The little pile of rocks was built especially to obscure the small opening of the little hide so that no one approaching would know it was there. I knew that this would have been my choice if I were in his place. It was in such a position that no one could easily find it, much less think that anyone could hide there. From its vantage point, Nails could easily see all approaches from the surrounding area as well as the rubble compound itself. Though I couldn't confirm much more, I had to assume that there were several escape routes throughout the little cave-and-building system. It would be a difficult, if not impossible, approach. I had to find a route that would take me closer without exposing me to the surrounding mountains or the artificial cave itself. I also could not afford to wait much longer. The shadows were lengthening and that would put Nails on higher alert, for that would be the time he would expect me to move. I had to keep doing the unexpected, moving through the impossible and doing the improbable.

I began to inch forward, flowing through the deepest recesses of the rubble and taking extreme care not to peer over the top of anything. Several times I had to come to an abrupt, nearly panic-stricken stop as faint concentric rings of nature would erupt from the rubble around me or the mountains above. At these times I had to define the maker of each disturbance before I moved on. Some of these disturbances were hard to define because they were so far in the distance. There was one disturbance in particular that I could not fully define. I knew that it was a predator, or several predators together, like hunting coyotes or dogs, yet it could also have been that kind of disturbance made by a man. Whatever it was it had a lethal effect on the land and in the core of my very soul. The only good thing about it was that it originated from far up in the mountains, and if it was Nails there would be no way he could see me moving forward on his suspected encampment.

Nearing the opening of this small cavern I decided to go to the extreme. Instead of staying within the deep crevasses of the rubble, I decided to take the most open and exposed route I could find. Again, considering what Nails would expect, it would be totally unexpected and probably overlooked. As I neared the vast open area I spotted several more tracks belonging to Nails. From his route along the outskirts of the area I could see he was using the same deceptive techniques I had taught him, but not at the level of the extreme approach I was using. There was no doubt that he chose this route to keep to moderate cover while inspecting the fallen buildings and rubble. Several times along the way I found his tracks looking back from where he had come and a few times he paused and hid for a few moments. It wasn't that he was hiding from anything or anyone in particular, but hiding so that he could watch and listen before moving on and into the little cave that lay not far ahead.

I could see and feel part of myself moving within his tracks.

After all, it was my training that was now guiding his move-ments, and thus he became part of me as I had to become part of him. It was so difficult to see part of myself in the tracks of this killer. But it was a relief to see him moving in this way too, because he was not really moving the way I would have, given the same circumstances. I would have taken a more obscure route, taking greater care to cover my tracks, especially during that time of night he had moved. So too would I have taken the route I was now taking; he had chosen to take a route that would throw off most searchers and hunters, but not me. I didn't know if he had done this to make me think he did not expect me to be following him, or if he did it to give me a false sense of security. As usual, these questions burned in my mind, making decisions difficult. If he did expect me and the tracks were intentional, then he might be luring me into a trap. I could not chance letting down my guard and underestimating his ability.

I decided then that I would assume the worst, without any doubt that he had set me up by deliberately setting down this type of trail and route of approach. Grandfather so often told me that the Scout always assumes the worst-case scenario, never giving in to what he thinks is an easy or safe situation. Otherwise it would be far too easy to underestimate the hunted, especially if the hunted had nearly the same level of skill as the hunter. By making this decision the going would be slow but safer. I knew that I had given Nails only a small percent of my skill and knowledge, but I could not know how much he had learned on his own. Experi-ence is the best teacher and he had vast experience as a warrior and assassin. Yet it is not enough to assume the worst, nor enough for me to become Nails, for now I had to hunt him as if I were really hunting myself. Even then I felt that this was not good enough, for something inside me demanded more. I had to hunt him at an even higher level than I would hunt myself; for once a decision was made that could lead to life or death, I had to hunt him as if I were actually hunting Grandfather.

As I approached the edge of the open area I watched both the surrounding mountains and the fallen structures very carefully. Every sense now had to be on its highest level of alert, well beyond the physical awareness that most people would consider normal. I had to pay rapt attention to every detail, even the faintest and farthest concentric rings had to be analyzed. Even the smallest and most insignificant nuance of nature missed or misread could mean death. I also had to find a route across the barren field that would not allow easy detection but would afford me the best possible scenario for counter-tracking. What made the journey worse was that I had to do it quickly, before the shadows grew longer. Moving too fast, however, could be as deadly as moving too slowly. It had to be a balance between the two extremes, decided more by experience and instinct than rational thought. Any thinking at this time would be a costly mistake as far as I was concerned.

The journey across the open area seemed like it took an eternity to complete. The area was so open that for most of the trip I had to rely on my camouflage completely. I had to trust fully my ability to become invisible, not only to Nails but also to any animal that I might frighten, thus sending vast concentric rings rippling across the valley that would be picked up by Nails. I had to force myself to stop thinking that Nails was up in the surrounding mountains too. There was an equal chance that the unknown disturbance I had failed to identify may or may not have been thrown by Nails. I didn't want to lean one way or another, at least not until the tracks told me which way to believe. I don't know if it was just dumb luck or spiritual intervention, but by the exact time I reached the cover of the far side of the field two things happened. Another ring came thundering down off the mountain, thrown up by a running deer, and definitely caused by man, for no other predator would send a deer fleeing this way. At the exact moment of the disturbance I found the freshest set of tracks belonging to Nails. The tracks led up the mountain.

The concentric rings had originated from near the midway point up the mountain slope, and given the identity and age of the tracks, it had to be Nails who set off those concentric rings. I also knew now that Nails must be on full alert. Sending out concentric rings of that magnitude is a huge blunder whether you are the hunted or hunter, and now Nails was left to intensify his level of awareness. And given his heightened state of awareness, it made movement for me much more difficult. So too I had to wonder if the deer might have been intentionally sent to flight as a diversion for me. Nails could have planted a nonlethal trap up on the slope on a deer trail that would have frightened the deer in this manner, leading me to mistakenly assume he was at midslope. He could now very easily have been lying in wait somewhere in the rubble; after all, I would have used the same tactic if

I were being hunted by him. Now movement and tracking no longer became an issue, for I was frozen in time and place, unable to make any decision.

I had to do now what Nails was probably doing, whether he intentionally threw the concentric rings or blundered into the deer accidentally; I had to wait and wait at the highest level of awareness. To do anything else at this point would be a deadly mistake. Now it had become a waiting game, a game to see who would move first, and the first to move would give away his position to the other. The only thing I had going for me was that Nails would not be sure where I was hiding, if he even suspected that I was in the area at all. However, I could assume nothing, for the concentric rings could also have been caused by a large predator like a mountain lion. The concentric rings thrown by a mountain lion and a man are distinctly different, yet at that great distance and far up the slope I could have misinterpreted their origins. There was also the outside chance that another man had thrown the deer into this panic. Though there were no signs or tracks to back this theory up, I could not rule it out entirely. After all, several times during the past few days a helicopter had passed far overhead.

This third-man issue began to haunt me as I waited motionless in the shadows of the rubble. The deer had definitely been sent into a panicked flight by either a man or a mountain lion. Yet if it was a man, and a possible third man that came on the scene, given the nuances of the concentric rings of the running deer the third man had to be good in the wilderness. I began to be haunted by the distinct possibility that the helicopter that had passed me several times over the past few days may have something to do with this third-man theory. I had assumed that the helicopter had been too high to see me, but if they were in fact looking for me they might have picked up my location after all. In the past I had trained many military helicopter pilots to track from the air, and this might be one of them, now come back to haunt me. This would explain the third man and the third man could be a sniper, sent to assist me covertly to make sure Nails was taken out.

Now the situation was growing very complicated. If there was a third man in this deadly game and he was a sniper, which I was growing closer to believing, then all of the rules had changed. Surely, Nails would also have picked it all up. Questions bordering on paranoia began to flood my mind. Was there a sniper dropped over the ridge by the helicopter? Was Nails aware of me there as well as the sniper, or was he totally unaware of anything at all? Was the sniper, if there was one at all, there to protect me, or did he just want to take Nails in dead or alive? And if I died in this deadly mix would I become nothing more than collateral damage to everyone concerned? After all, for the past few years I had not been on the best of terms with the government men I had been training. I didn't like teaching Grandfather's skills to make people more efficient killers and I let them know it, backing away from the teaching process as much as possible.

I was frozen in a state of mental paralysis. Fortunately I realized what was going on within me. The extreme fatigue, dehydration, and the relentless cat-and-mouse game with Nails and the possibility of death had caused me to second-guess myself. It is essential, on the edge of physical and mental extremes, that a Tracker keep cool and objective. Irrational thinking could be deadly, and assuming that misread concentric rings could in fact be a sniper was a stretch of the imagination, more hallucination than reality. Though I could not fully dismiss this third-man theory completely I had to set it aside and make it only a distant possibility. I could not allow the distraction of another person in the mix to detract from my focus on tracking Nails. Splitting my attention, especially where Nails was concerned, could be a deadly mistake. No doubt, if there was a third man, and even if he was a sniper, I would probably not be the target. After all, it was Nails that they wanted, not me. However, I could not afford to blow this theory off completely, only keep its possibility in the back of my mind. To ignore it altogether would be a mistake. A Scout always assumes the worst.

I refocused my attention on the tracks that lay before me. At least I could study them and determine what time Nails had headed up the face of the mountain, without being seen. This would break the monotony of the intense waiting and help clear my mind, all the time resting before I headed to the slope. From the cover of this place I would also be able to hear any movement up on the slope. These tracks were also a gift because they clung close to cover and I could follow them quite a way without revealing my position. I just had to be careful that my attention stayed split between reading Nails's tracks, remaining hidden, and listening intently for any subsequent movement. Unfortunately as I followed the tracks I began to get sucked into them. The tracks indicated that Nails was doing something odd, though I could not read them easily. The pressure releases had been partially erased and it was like reading text with every other word missing. It demanded my full attention, but I had no idea what the tracks were really telling me.

The one major flaw Grandfather was always trying to correct in me as a tracker was the way I would get sucked into reading a track. It wasn't so much the reading but the way I would block out everything else outside of the track. So many times I would be tracking a deer and not notice that at the point of reading the track another deer was passing by in front of me. This would always give Grandfather a good laugh, for I would miss so much of what was going on around me. He was constantly telling me, "Vary your vision, Grandson, vary your vision." That was one of my major downfalls as a tracker, for a good tracker is paying attention not only to the track but to the universe around him. It got so bad that as I tracked Grandfather I would pass right by him following his tracks, only to be kicked in the ass by him standing just inches from the track I was gazing upon. As I grew older, Grandfather no longer found this habit amusing. He would often tell me that such a mistake, such focused absorption, could cost me my life or the life of another. I thought that I had outgrown this bad habit.

The events that followed are now only a blur, an instant in time and reaction, but a heartbeat from eternity. As I crouched over one of Nails's tracks, trying to read what it was telling me, a

chain of events occurred, almost at once. It was not the knowledge of what the track was telling me that I could have articu-
lated at the time but rather the instinctual reaction to it. If I had
remained crouched over the track, the bullet would have hit me
in the back of the head. If I had just stood, the bullet would have
hit me in the center of my back and crippled me from the waist
down. But in a blinding reaction, I stood and turned hard to my
right. It was at the exact moment of realization of what the track
was telling me that I reacted like a striking rattlesnake. At that
moment I realized that the track was telling me that Nails knew
that I was tracking him and I clearly heard Grandfather's voice
demand, "Vary your vision." At the same time I also knew that he
was behind me and that the situation was lethal. I just reacted
without thought and with every ounce of fear, rage, and power I
had within me.

I remember feeling the bullet hit me, but not hard. It was
more a burning pain than anything else and not debilitating, but
it unleashed the animal within me. My life was in danger and all
I could do was to react with violence, more violence than I have
ever known. I hit Nails hard, so hard that I could feel the bones
breaking in my hands as I connected with his face. He went
down in a cloud of dust and I crumpled beside him. Thoughts
swirled through my mind. I was spent. I exploded with the last
energy I had and survived. I was so angry at myself for making
such a stupid mistake, a mistake I had made over and over again.
I should have known immediately that the tracks were telling me
that Nails knew I was on his trail. I should have known that they

were laid down by Nails to divert my attention, forcing me to think he was far up the slope. I was humiliated because Nails had beaten me. Being beaten by one of my students was more than I could bear. Worse yet was being beaten by myself and a stupid mistake that could have cost me my life.

The winds of the approaching storm began to kick up dust and the sky blackened. Soon it would be raining and I thought about dragging Nails to the cover of a small recess in a nearby building. I wondered when he knew that I was tracking him. None of the tracks up until this point had indicated that he knew anything at all. It was only these tracks that showed any indication that he knew that I was tracking him. I wondered when I had made my mistake, when I was seen, or if the tracks were just laid down out of Nails's paranoia, where he only assumed that I was tracking him. Whatever the reason and purpose, he was laying in wait for me and I had passed right by him, just as I had passed by Grandfather so many times before. Only this time I was not kicked in the ass but shot in the back. There was no doubt in my mind that I would not make this mistake again, for it will forever be etched in my very soul.

Nails still hadn't moved and I began to fear that he was badly hurt, or worse. I began to crawl closer to him now, thoroughly enraged at myself for that damn mistake. I had been hoping to find him asleep and capture him after removing his gun or guns, but now all that had changed. Because of my incompetence both of us were hurt and out in the middle of the desert. I knew that it would take several days to get us out of there, and given the extent of the injuries it would become a life-and-death struggle. Even in the best of shape, crossing the desert and relying only on survival skills would be difficult. The only consolation, I thought, was that the oncoming storm would at least provide us some much-needed water. I began to pray and pray hard, for my body and mind were spent and I had hardly enough energy to sit up. It was a struggle just to roll from lying on the ground to the sitting position, but at least the pain of the gunshot wound had diminished and the bleeding had stopped.

As I eventually came to a painful sitting position I gasped in amazement. Not a hundred yards away I saw a man approaching. He was wearing some sort of camouflaged uniform and carrying a high-powered rifle. He carried the rifle with its muzzle up, indicating that he was not about to shoot, neither was he concerned that we were any threat. I could feel my face burn with the rush of angry blood, for I had made another mistake. I had discounted another presence, the third man, and here he stood in bold reality. I was humiliated again, knowing damn full well that I had second-guessed my second-guessing and had nearly written off all of the concentric rings. It was obvious now that the helicopter had spotted me and he was there to track me tracking Nails. I should have known. I had only tried to hunt and stalk Nails, never realizing that I was being set up and followed. I should have also disappeared altogether. That third man could have become a liability by alerting Nails earlier in the struggle. I wonder if in fact he was the reason Nails had known that I was following him; after all, Nails would know how these folks operate.

As this man walked toward us I felt a strange kinship with Nails. It was almost the feeling that I was somehow on his side in a macabre sort of way. After all, it was he and I who had played this deadly game of hunter and hunted. It was I who had trained Nails, and in a way I was responsible for him. It was we who survived this desert and all of its hardships, and not this outsider, this third party with the rifle. Anger burned within me as I watched the man's approach. I connected him to all the things that had gone wrong, the circumstances that had created and driven Nails to do what he had done. That man just spoke volumes of what was wrong with the way people are used up and then thrown away. For a brief moment, as he drew nearer, I thought about taking him out and letting Nails go, but at the same time I realized that this was coming from a place of anger, fatigue, and my failure as a tracker. All I could think about was how embarrassing this would have been for Grandfather.

The man walked up to us and surveyed the situation for a long moment. I could not even speak, for it had been days since I had uttered a word, and I was also afraid of what I might say. He first glanced at Nails with a look of disgust and loathing and then at me. I could see from the makeshift sort of uniform that he was not from any military outfit I knew, though he was carrying a U.S.-made .308 caliber sniper's rifle. After a long and thoughtful pause he said, in very broken English, "Mr. Brown, I have come to take possession of your prisoner." His voice was very formal yet, ironically, very passionate toward me, which stunned me for a moment. I tried to question him but nothing would come out. My throat felt parched and welded shut. My brain could not fathom what he was saying and I could not connect any words to my thoughts. Without waiting for me to answer, he said, "Mr. Frank has asked me to retrieve you and the prisoner and bring you both back to our compound." With that he spoke into his radio and in a few minutes a pickup truck came up the rutted road.

As I painfully got into the front seat of the pickup truck, too tired to argue the circumstances, I watched as two other men, dressed the same way as the sniper, loaded Nails into the canvas-covered back part of the pickup bed. He was placed on an old Korean War–era stretcher, but loaded as if he were just a pile of garbage. I tried to protest his treatment but could not muster the energy. As we drove away I looked through the back window of the truck, through a rent in the curtain, and saw that they were not doing anything for Nails to help him. The guards just sat there and talked. I on the other hand sipped fresh water and tried to relax for the long ride out. I had enough to do to fight fatigue, the nagging pain in my back, and to try to figure out where we were, all the while desperately piecing together all the events of the past several days. What haunted me the most was reliving all of the horrible mistakes I had made, not to mention the biggest mistake of all—that of teaching people like Nails in the first place. Some-place in the middle of nowhere along this rutted excuse for a road I gave into a much-needed and very profound sleep.

Next thing I knew I was abruptly awakened by the truck pulling into an old compound of sorts. It was some sort of official-looking building, still in the middle of the desert and with a small town built up around it. As I was ushered quickly into the building I noticed the Suburban I had ridden in a few days before, but Frank was nowhere to be found. I glanced down the driveway and spotted his tracks along with the tracks of several other individuals like him. I was taken into a dank and dusty doctor's office that looked more like a backwoods veterinary clinic and sat on an old table covered by an equally old mattress and worn sheet. There I met a frail old man in a dingy white lab coat sporting an ancient stethoscope, and I immediately assumed he was some sort of doctor. Applied to him the word "doctor" was a stretch, for his hands looked like they hadn't been washed since he plowed the back field. Yet when he spoke, though in broken English, he sounded both compassionate and knowledgeable, which alleviated any fears I had.

Reluctantly I allowed him to look at my back, but he seemed to be rather unconcerned about the wound. Apparently it felt worse than it actually was. He jabbered about how lucky I was that it was only a graze; even though the bullet had passed into my back it only threaded its way under the skin and out again like a splinter. He told me that he was not going to stitch it up, but would leave both the entrance and exit wounds open so that they would drain out any infection. At least the bandages he applied were sterile, but the syringe he used to inject the penicillin was a little suspect. I thought it looked like the last time it had been used was to inoculate livestock. At least now, given the water, the sleep, and the bandaged wound I began to feel better. Questions began to surface and I asked the doctor how Nails was doing. He grew very quiet and simply told me that he would find out. With that, he jumped up and abruptly left the room. I did not like the way he reacted to my question and began pacing impatiently as I awaited his return, which seemed to take forever.

Finally after what seemed to be the better part of an hour another man came in wearing the same strange military-type uniform as the gunman. In far better English he told me that my prisoner had been turned over to Mr. Frank and that my services would no longer be required. I pounded him with questions but got no answers. He kept saying that my prisoner had been turned over to Frank and the proper authorities. The more questions I asked about the situation the more evasive the man became with his answers. I grew very angry and exasperated, for it was apparent that someone was trying to cover up something and that was pissing me off. It infuriated me that I did all the dirty work and put my life on the line, and then was cast aside like an old rag. "Your services are no longer required," what a crock of crap. Yet it should not have surprised me, for this same thing has happened so many times before, and each time I allow myself to get angry and upset over the evasiveness.

After another hour of waiting I was rushed out of the building and into another pickup truck. The driver spoke no English at all and the two men that rode in the back bed were not the same ones that rode with me back from the desert. It was obvious that the doorway to answers had effectively slammed shut and I was on my way someplace in a hurry. As we drove I could see a large town for ahead and it was obvious that we were going there. At this point I was expecting to eventually be put up in some motel and be debriefed by Frank, but to my surprise that did not happen. Instead they took me directly to a small airstrip and a small twin-engine airplane. Given the scant markings of the plane I could not tell the country of origin or even if it could fly. It looked so old and weather-beaten I could no longer tell what color it had been painted originally. Even the seats were frayed and worn. As I was rushed into the passenger seat I realized that they wanted me out of there as quickly as possible.

I tried to protest but was met with blank stares. I was livid with the rushed treatment, even though my escort tried to remain pleasant and cordial. The angrier I became the more official they tried to become. I was so pissed off that no one was telling me anything. I even made the comment that I was going to go home and go directly to the *New York Times* with my story, but they didn't seem to care. For that matter I wasn't even sure they understood anything I was saying. Within moments we were rolling down the runway, my escort making damn sure that I took off. It also became obvious that the pilot did not speak English either so I didn't bother to ask him any questions. In fact the pilot wore an expression that said he did not want to be bothered at all and he did not speak a word throughout the whole flight. Eventually, after hours of flying over ever-changing landscapes, we landed at a major airport, where I was immediately ushered onto another plane, this time a jet, but with no markings other than the required numbers. To my amazement this new plane was nearly empty except for a handful of official-looking people.

As I grabbed some water and a sandwich at the galley and made my way to my seat I was met by a man who was dressed the same way as Frank. He smiled at me warmly and said, "Thank you, Tom." Before I could get a word out, he continued, saying, "You are well aware of the routine. There is no debriefing. Just go back to your life. The fewer questions you ask the better . . . and once again, thank you." With that he turned and walked off the plane, giving only an obligatory wave as he left. Immediately the door was closed and almost as quickly we were under way. I was enraged, to say the very least. I was angry at the system, angry at being kept in the dark, and angry over what I'd just lived through. Most of all I was angry at myself and all of the mistakes I had made, mistakes that could have cost me my life. What bothered me most is what happened to Nails, and that they rushed me off so fast, without so much as a debriefing. Something was wrong with this whole mess. I wanted answers desperately, but I knew from past experience that there would be none.

I felt much like Nails in a way. I had been used and then abandoned when I was no longer needed. I wondered how many operations and missions he found himself on that were so clouded by questions. I wondered if he ever knew that he was being used, only to be abandoned and denied, eventually only to be hunted down like a rabid animal. It troubled me that I did not know where he was, how badly he was injured, or if they might just have killed him outright, thus avoiding any courts. The most haunting and disturbing factor of it all was the sickening feeling that I might not have just injured or maimed Nails, but might have actually killed him. That I may never know. Even what few friends he had have not heard from him since the murders occurred. Now, not only will I have to live with the ghost of Nails, but with all the horrible mistakes I made when I tracked him. Not knowing is one of the greatest tracking mysteries I will face for the rest of my life. Someday, somehow, like all trackers, I will find answers.

THE EYE OF THE TRACKER

THERE IS A way that a Tracker looks at the world that is far different from what most people can even imagine. As I've said before, it's called the Tracker Point of View, a way of not only seeing, but hearing, smelling, tasting, and touching the environment that transcends the normal limitations of human sensory awareness. Yet it goes well beyond those heightened human senses and enters the deeper world of spiritual understanding through the vehicle of Inner Vision. Some may call this Inner Vision a sixth sense, imagining that it is some sort of spiritual gift given to only a select few individuals and denied the rest of the people, but this is not true. The Tracker Point of View and this Inner Vision can be easily taught to anyone and just as quickly learned. To learn this process, this way of looking at the world through the eyes of a Tracker, one needs only an intense interest, a passion, and dedication to constant practice.

The Tracker Point of View and the vehicle of Inner Vision is not something that is turned on and off at will, but something that is constant, always in place, and a habit that cannot be broken. My students often ask me, When do I practice Tracking, practice the Tracker Point of View, and use Inner Vision? That should not be the question they ask. Instead, the question should be, When am I not tracking? When am I not using the Tracker Point of View and the Inner Vision that is so much part of a Tracker's life? The answer would, of course, be "never." To be a Tracker, at least as Grandfather defined the philosophy of being a Tracker, is a lifestyle, or rather, a way of life. But Grandfather also defined a Tracker as something more than just someone who follows footprints.

Grandfather did not separate Tracking from Awareness, for to him they were one and the same. One could not exist without the other. One could not be a good Tracker without being highly aware, nor could one be aware without being a great Tracker. Grandfather called survival "the doorway to the Earth," but he called Awareness and Tracking "the doorway to the spirit." That is the awesome power that he gave to the philosophy of being a Tracker and all of the skills that go with that philosophy. Grandfather often said that Tracking and Awareness were the most important physical skills a person could possess. Subsequently, every day without fail there would be lessons in both Tracking and Awareness. Still, and most important of all, were the spiritual skills. In the final analysis, the spiritual skills and philosophies were emphasized more than anything else that he taught.

As I think back over all of the years and teachings of Grandfather, I suspect that he not only considered Tracking and Awareness to be one and the same, but so too did he believe that the philosophy of spirit was also part of that whole concept. Though he did not come right out and say that Tracking and Awareness could not be learned and understood without the way of spirit, it was obvious through his teachings that Tracking, Awareness, and Inner Vision were all intricately related and the connection between them could never be broken. After all, Grandfather called Inner Vision "the voice of the spirit," and Inner Vision was the most important skill a Tracker could possess. This relentless passion for Tracking, Awareness, and Inner Vision became our main focus, our way of life, and our window to the world where Grandfather lived. This Tracker Point of View was in actuality seeing life through Grandfather's eyes.

Living life through this Tracker Point of View, through the Eye of the Tracker, became the single most important driving force in our lives. It opened worlds to us that few dream of ever seeing or experiencing. Looking through the Eye of the Tracker was practiced so much that it eventually became habitual, a way of life that could not be broken or turned off. It became so much a part of our makeup that we could not view life in any other way, nor would we want to. As was common with Grandfather, we learned to see in a glance what would take most people days to discover, if they could discover anything at all. However, there was also a downside to this Eye of the Tracker way of experiencing the world. We would not only see all the wonders and mysteries of the world around us, but we would also see the negative and the ugliness.

Equally as important to a Tracker as Awareness and Inner Vision was the constant quest to answer what Grandfather called the Sacred Question. The Sacred Question is the driving force behind all of what it means to be a Tracker, all the fire and passion it means to be Aware. The Sacred Question is the continual asking of the question What happened here? or What is this teaching me? The search for answers to the Sacred Question demands keen analysis, constant awareness, close observation, and experimentation. It defines the intense need to know and understand all of the things around us found both in the natural world and the worlds of man. Grandfather so often stated that the biggest problem with the Sacred Question was the failure to ask it. Essentially, what he was saying is that, like Tracking and Awareness, the asking of the Sacred Question should be a constant, a way of life, where it is never turned off.

The Sacred Question to me encompasses everything, from the smallest and most mundane questions to the larger questions of life and Tracking. At one moment I could be studying why a person puts out a cigarette butt in a certain way and in the next minute trying to discover when the last time was that an animal drank water as defined by its tracks. To a Tracker, the quest to define and answer the Sacred Question never grows tiring or boring, but rather evokes the deepest passion and drive. I find it just as interesting studying the flatness of a fleck of dust on a stone that depicts a mouse toe print as solving a murder mystery through the way a speck of blood has been transferred many yards from an actual crime scene. It makes little difference where I am or what I am doing, for the quest for the answers to the Sacred Question are always with me, always a constant in my life. And life is a series of unending mysteries, begging to be solved, answered only by the total dedication to the Sacred Question.

This Eye of the Tracker, defined as the equal balance of Tracking, Awareness, Inner Vision, and the quest to answer the Sacred Question, is what it takes to be a Tracker. It is a lifestyle, a way of life that is never turned off. It is the philosophy that drives the Tracker to reach ever greater heights of understanding and the passion for solving mysteries. In my opening lecture of the basic Standard Class at my Tracker School I always tell my students that my main ambition in teaching the class is to teach them to "see through Grandfather's eyes." To me, that is the most powerful statement I can make, the most important skill I can give to them, for it opens up a world that is hidden to most everyone. Yet I also try to instill in my students that to see through Grandfather's eyes is a way of life, a journey and an adventure that never ends. This quest, this journey to see through the Eye of the Tracker, is the Vision, and as Grandfather so often said, "It is not the Vision that is important, but the journey."

All Trackers are on this same journey of understanding, no matter what their skill level. As both an instructor and Tracker, however, the biggest problem I see in students is that they try to turn this journey, the quest for understanding, on and off instead of living it constantly. I so often say to my students, "I can give you the skills you need to become a Tracker, but I can't give you my passion or experience." Only they provide themselves with the passion, experience, and the drive to become a Tracker. Seeing through the Eye of the Tracker is the religious dedication to live Awareness, Tracking, Inner Vision, and the quest to answer the Sacred Question every moment of every day. Unfortunately I have to warn them that once they adopt this way of life they will not only have a beautiful new world open to them, but they will also bear witness to all that is ugly. To truly see through the Eye of the Tracker, we cannot choose only to see the once-hidden worlds of beauty, but also have to see all that is negative, ugly, and defiled.

With this lifestyle of the Tracker in mind, there are a series of Tracking cases in my Tracker Case Files that I call Accidental Tracking Cases, meaning Tracking cases that I stumbled upon and subsequently solved because I live the Eye of the Tracker all the time. Certainly, some of these Accidental Tracking Cases I was officially called in to solve, but others I stumbled upon, not knowing what I was getting into or even if it was a Tracking situation at all, until the case was solved. Living the Eye of the Tracker does not depend on environment or situation, for it is constant. The Eye of the Tracker is equally open in the city and suburbs as it is in the wilderness. It does not choose when and where, but is always. What follows is a small collection of these

Accidental Tracking Cases, many of which I stumbled upon because of the blessing and curse of the Eye of the Tracker.

MISSING PERSON

I was driving along the interstate, in no rush to get anyplace in particular. Most people would call it a road trip, but my usual road trips have neither time nor destination. That is the beauty of being a survivalist, for being able to survive anyplace without anything is the ultimate freedom, a freedom that very few people would ever know or can even imagine. With survival I can go anyplace without any luggage or equipment and just the clothes on my back, able to stay away as long as I want and never needing anything from society. On this occasion I had chosen to drive rather than walk or hitchhike, which was my normal means of getting across the country. Just driving a vehicle seems so fast, so removed from the landscape, and so much is missed. A vehicle is nothing more than another piece of equipment that has to be taken care of and limits my access to the deepest wilderness areas. The only time I used to drive was when I had a particular place in mind that I wanted to visit and didn't want to take the time to walk there. This was one of those rare occasions.

Driving for me was always difficult for many other reasons. It was bad enough that I had to find a place to hide the car when I did decide to leave the road and wander off, but having to feed it fuel, stick to roadways, and make repairs seemed at times more trouble than it was worth. My friends used to laugh at my definition of a roadside rest area or restaurant. Unlike most travelers I did not gauge my stops by how far the next gas station, restaurant, or hotel was located, but decided where and when to stop to eat or sleep using my own set of rules. My place to pull over and sleep for the night was determined by where I could hide my vehicle from passing patrol cars or nearby towns and houses. I would stop to eat when the roadkill looked fresh and there was a good assortment of wild edible plants nearby.

Another problem with driving for me was what I drove. At the time I had an old, very old, Toyota Land Cruiser, the old box-style body type. I had bought the thing very well used and I suspect that it was at least on its second or third engine and transmission. At one time it had a hard top, but I had taken that off long ago for two reasons. One reason is that the original hard top had only one side window that was not broken and the frame had rusted through so much that it was in danger of blowing off the vehicle anytime my speed exceeded fifty miles per hour, though I doubt the truck could have really done that speed, and I couldn't know anyway since the speedometer did not work. The second reason it had no top was that I hated to be boxed in. Summer or winter, rain or snow, I would drive without the top. I loved it. After all, if I had to drive I wanted to be able to see the landscape and the sky without any rusty restriction. The Land Cruiser was in such bad shape that I used to gauge my trips not by miles per gallon but parts per mile, and I carried a grand assortment of spare used parts wherever I went.

Yet the biggest problem of driving for me was not so much

that I was a bad driver, but a Tracker. My driving habits, to the untrained eye, were much like someone who had a blood alcohol level exceeding all human capacity. I know I must have pissed off many a police officer who pulled me over thinking I was intoxicated, only to find out that I could walk a straight line or recite the alphabet backward. You see, driving and being a Tracker do not really mix. I would try to drive, keeping one eye on the dotted line and the other along the roadside searching for signs of animals, interesting plants, trails, rocks, or anything else of interest. Unfortunately for me watching the fringe areas along the roadside used to win out over watching the road. Speed was also a variable factor, depending on what interesting thing I was passing, what hawk might have just killed a rabbit on the median strip, or what skull from a roadkill could be collected. It was not uncommon for me to slam on my breaks in the middle of the road, swerve onto the shoulder, then dash out and pluck an animal skull or feather off the side of the road. Other drivers would not stay behind me for very long.

I also had the nasty habit of hanging partially rotting animal skulls from various parts of my truck. If they were too rancid and maggot-ridden to put inside the vehicle I would simply tie them to my bumpers and let the elements finish the rotting process through the rest of my journey. I often wondered why my truck was never stolen or the contents were never taken when I was off in the bush exploring, but after hitchhikers refused to accept rides from me on various occasions I realized I had one of the best antitheft devices that could be found. Not only was my old truck a piece of beloved junk, but it always had an assortment of various bones, skulls, chunks of wood, rocks, and other flotsam hanging from it at any given time. Sometimes the smell of rotted flesh became so bad, especially on hot summer afternoons, I would have to run to the truck while holding my breath and pray it would start and get up to speed before I had to breathe again. Another blessing was that I was rarely ever tailgated, especially in heavy slow traffic.

Being a Tracker, a survivalist, and having been raised in the woods, all of this was totally natural to me. At the time I couldn't understand why a person would be sickened by a partially rotting deer skull hanging from my rear bumper or spit-cooking a road-killed groundhog along the interstate. After all, my nature collection was expansive and growing all the time, and collecting things that are found dead along the road is just another way of adding to my collection. Collecting good rocks, pieces of firewood, plants, clay, and other things was like shopping. Even though I was driving I still had to survive and collect all of the tools needed for survival along the way. To this day some of the finest stone arrow points I've made were formed from rocks I found while driving throughout the country. Needless to say, the Eye of the Tracker and all of the philosophy that comes with it does not end when I enter a vehicle, which unfortunately still holds true today.

So, on this particular day, driving along the interstate without time or destination and traveling in my old Land Cruiser, something strange caught my eye. It was not so much that it was strange, but it just did not make sense. A particular set of tire prints from a pickup truck had skidded through the grass along the roadway, crashed through brush that had nearly sprung back into place, and disappeared down a partially hidden embankment. Normally this would not puzzle me because I see these types of tracks all the time, but what did bother me was that there was no evidence of the pickup truck exiting. The truck, given the damage to the grasses and the brush, had been moving at a high rate of speed, and the lack of any exit evidence really began to bother me. I slowed down for a bit to decide whether I should go back and investigate but gave it up. I began to think that there might be a hidden off-road trail on the other side of the brush that I had failed to notice and the truck driver might have decided to make a quick exit, especially if he were trying to hide from a patrol car.

I'm not going to incriminate myself and say that I have used the same tactics, but I certainly know people who have, especially in the Pine Barrens of my home state. It was not uncommon for a speeding 4X4 truck to run off the road just to avoid a closing patrol car and avoid a speeding ticket. I began to think that it was possible that they were doing the same thing in other states, and especially along this interstate. Still the tracks gnawed at me for two reasons: the high rate of speed at which the tracks indicated the truck had run off the road and the landscape indication of a rather steep drop-off on the other side of the brush. The more I tried to explain away the tracks, the more the reality of the tracks and my Inner Vision told me different. This internal struggle finally got the best of me and I pulled off the road at a rest stop about thirty miles west of the tracks to think it through.

I went into the restroom, which was a rather typical and quite primitive cement structure, not much better than an outhouse. I also knew that many of these stops had roadmaps posted on the walls to aid travelers. Since I didn't make a habit of taking a roadmap with me I thought this might afford me a chance to look at the map and see if there were any linking roadways paralleling the interstate thirty miles back that the pickup truck might have taken. Unfortunately as I entered what was supposed to be a small lobby, with its usual assortment of snack machines, I found no map I could use. The one posted on the wall was so old and weather-stained that it was unusable. It also covered most of two other states, rendering the reading of any backroads next to impossible. I gave up on trying to read the map, went to the bathroom, and was about to leave the building when a poster on the wall caught my eye. Seeing the poster pissed me off at first because it was taped to the wall right next to the map and I hadn't noticed it.

The poster was handwritten in block letters and at the top of it were the words "missing person." Those words had been the ones that caught my eye, not only because that is what a tracker does, find lost people, but because it also showed a certain degree of desperation. Judging from the way the poster was made and what was written, it had come from a distraught family. Apparently a seventeen-year-old boy had disappeared six days ago, last seen at his girlfriend's house late on a Tuesday night. It went on to say that he was driving a 1969 Ford pickup truck, which was dark green. He had left his girlfriend's house, which, after looking at the faded map, I discovered was located about thirty-five miles back along the interstate. I began to piece things together in my mind and heart that I had tried to brush off many miles back. I realized why my Inner Vision had kept those tracks gnawing at me.

I turned the old Land Cruiser around on the median strip and headed back to the area where I thought the tracks might be located. At the time I discovered them I did not pay much attention to the mile markers or the nearby exits, so I had to travel more by landscape than by signs or mile markers. I began to race the clock, fearing I would run out of daylight before I made it back to the area, so I had to force the Land Cruiser to the most speed I could get out of it, which was still below the legal limit. The old truck was just too tired to do much better. I decided that what would be best would be to exit at the town his girlfriend lived in and head back west, which would give me a better chance of locating the tracks again. Even though this might cost me much-needed time and miles, I could not afford to take an early turn and undershoot the track location. I had to locate the tracks before full dark because I didn't want to search with just my weak headlights. I knew that if I missed the tracks it would be best to spend the night someplace close and search again in the morning.

I took the exit to the girlfriend's town and reentered the interstate in just a few turns. This was a gift because some exits do not also have entrance ramps, but this one proved in my favor. There is nothing worse than trying to find an entrance ramp in a strange town, especially looking the way I looked and driving the thing I was driving. That course would always invite close scrutiny from the local authorities and I could not afford the time to be pulled over by a local cop. I knew also from past experience that it is very difficult to explain to some people the finer concepts of tracking, especially when the police are suspicious of me in the first place. That option would be out of the question. Though at that time in my life I was quite well known by law enforcement throughout many parts of the country for my tracking, this town was not one of them. As luck would also have it, I had plenty of daylight left to cover at least the next fifteen miles.

About six miles up from the entrance ramp I slowed my truck and began to search the roadside. Light was fading fast and I didn't want to overshoot the tracks. At the same time that I instinctively slowed, more by Inner Vision than tracks, the landscape and the hidden ravine began to grow familiar. It was not long before I fully pulled my truck off the road just in front of the tracks of the truck I had spotted earlier. With no time to waste I ran to the tracks and began to sprint toward the battered brush. To the untrained eye, the brush would not have looked very different from any other brush, but to me this brush was battered. As I had suspected from the lay of the landscape and the growth of the trees there was a ravine just on the other side of the brush. The tracks went right down the embankment and in the distance, even in the fading light, I could see the back end of a Ford pickup truck.

As I rushed down the hill and approached the truck I could clearly see that it had smashed into a huge old tree at the bottom of the ravine. The truck had come to rest about forty yards off the interstate, and where it lay it would possibly not even have been spotted in the winter when there were no leaves. I realized now why the boy had not yet been found. The crash site was just too camouflaged. As I drew near the truck the stench of rotting flesh became overwhelming. I could see the boy tangled in the wreckage, part of his body hidden under the steering column and compressed dashboard. I knew that he was dead so I didn't approach the scene any farther, as this would only make the gathering of evidence by police that much more difficult. As I scrambled back up the hill I began to consider my options. Should I flag down a passing motorist? That seemed out of the question because no one in their right mind would stop for me. Should I wait for a patrol car? That could take hours, I thought, for I had not seen a patrol car all afternoon. Just as I was thinking about heading back to the town to go for help I broke through the brush and to my amazement my problem was solved.

Behind my truck sat a state patrol car and the officer was standing next to my car on the passenger side, his back toward me. It is by habit and long practice that I always walk silently, so I could only imagine what went through his mind when I tapped him on the shoulder. As he spun around and looked at me, trying to go for his gun, sputtering out commands at me and visibly shaken, I just put up my hands and stood there and waited for him to calm down. The gun was shaking so badly in his hand that he couldn't have hit me with a shot if he tried. Without as much as a question he began to yell at me for "pissin' in the woods," stating that there was a rest area not far up the road as he eventually calmed down. I was about to tell him about the pickup truck when he began to tell me that he didn't like my looks, was going to write me up, and said that I was some sort of wacko, pointing back to my truck.

Finally when he calmed enough I told him that I was sorry for taking a piss, but I found a green Ford pickup truck down in the ravine. His jaw dropped and I could see the interest replace the anger and fear in his eyes. From many years of experience and given his initial attitude, I was not about to tell him about back-tracking the truck, or about tracking at all for that matter. Holstering his .38 special he raced back to his patrol car, where I saw him make a quick radio call and grab his flashlight. At first I think he was going to take the dash-mounted shotgun, probably to cover me, but had thought better of it when he looked back in my direction. He raced back to my truck and asked me to lead him to the pickup. I knew that he would never find it on his own, but I also knew that he felt more comfortable with me walking in front of him so he could keep an eye on me. I could sense that he was still a bit shaken up by my silent approach and my looks in general.

I could hear him stumbling through the brush and down the hill behind me. I knew from his walk that he was not that famil-iar with the bush, even though this area was considered deep country. He kept asking me how much farther it was, which had begun just as we passed through the brush. As he followed me farther down the hill he asked me why I had come into the woods so far just to take a piss. I told him simply that I was actually looking for a comfortable log to squat on to do a little more than just piss, and that seemed to satisfy him. I just didn't want to get into telling him about following tracks at all. I knew that this would get into lengthy police reports, and a full day had been taken out of my road trip already. As soon as I told him that the truck was just up ahead I could see his flashlight reflect in the taillights. He rushed past me, almost knocking me aside as he slipped by.

As he neared the crushed cab I asked him if there was anyone in it, playing dumb to the fact that I already knew. He hesitated and said, "I don't know yet," though I could hear him gag as he said those words. I could tell that the young cop had never encountered a rotting body before, possibly no bodies at all, and he was not ready for what he found. I could hear him spitting and gagging back up the hill, trying to regain his composure as he joined me. I could tell that he was uncomfortable, not only with me but with being in the woods and the stench of rotted flesh. I began to ask him about the truck, but he gave me very little information. This struck me as funny because I had no idea why he was trying to be so secretive about the truck. We waited on the side of the ravine, midway between the truck and the brush, apparently awaiting his backup. I could tell that he was torn between wanting to wait by the truck and waiting by the road. He kept flashing his light in the direction of the truck, as if to make sure it was still there.

He also seemed to not want me to go anywhere, in case he could not find the truck again. As we waited I asked about the truck again but he diverted his attention to me. He asked me what I was doing in the area, where I was going and where I was from, but he never asked me my name. In fact, he never even took out his notebook or asked to see any of my identification. He seemed nervous and edgy, totally out of place in this dark ravine. Finally a crackle came over his big walkie-talkie and he answered the call. He was apparently trying to direct the backup patrol to his location, but with little success. I simply told him to shine his light up into the treetops and they would see the light from the road. He looked dumbfounded at the idea but did it anyway. To his amazement the backup saw the light, and I could see lights now coming down the hill and heading toward us. He went up the hill, just out of earshot, and talked to the two other men.

The three cops headed back to me after a short conversation, and one of the backup officers said, "Thank you for your assistance, sir, you can go now." He was dressed in a higher-ranking uniform and was much older than the young officer I had first met. But he also had a no-nonsense attitude that implied trouble if I didn't leave the scene immediately. As I silently slipped up the hill I could clearly hear the first officer say, "Damn curiosity seekers! They always pull over when you don't need anyone around." That really pissed me off and I stood in the shadows for a while trying to pick up parts of their conversation. Though I couldn't hear the words it sounded more like they had just won some coveted prize and showed no remorse or pain for the dead boy. I was just leaving the scene when an ambulance, a small fire truck, and a wrecker arrived. I could see three lights illuminating the treetops, guiding the rescue team down the hill.

I wondered for the next week why the first officer had been so vague with his answers and why the older cop had asked me to leave the scene. Back then I never took any credit for assisting the police. In fact, I avoided all publicity. To me, Tracking is like a cherished gift and I gave that gift back by tracking. I didn't want the publicity, yet most law-enforcement agencies I assisted would say that "the Tracker" had helped them, out of respect for me and my ability. After all, I was always known to everyone as "the Tracker," never by the name Tom Brown. It infuriated me that I had been treated in this way, with at best a halfhearted thank-you, and then hearing the words "curiosity seeker" used to describe me.

What should have been evident the night of finding the pickup truck became clearly evident a week later when I drove back along the interstate headed home. Stopping in a local rest area near the scene I spotted a local weekly newspaper. The headlines read "Local State Police Officer Credited with Finding Missing Bernard Boy." I read some of the article that I could see through the glass window of the display box. It went on to say that the officer had noticed something "amiss" along the interstate and went on to find the missing boy and the pickup truck. I was enraged because as I read on it made him sound like some sort of hero with phrases like "scaling down a harrowing ravine in the dark of night" and "standing guard alone as he awaited backup." It really made me feel violated. But, as with everything else with publicity back then, I just let it go and let him bask in his phony limelight.

Many years later, however, I did get some small but sweet revenge. After my school was well established along with my reputation and the reputation of my Tracker Teams, I found myself teaching a special police class out in that part of the country. Wouldn't you know it, but the first officer was a member of the class. He had since been promoted several times and now was a captain. He had that certain "I am so good" attitude, but to me he was not that good. You see, he hadn't changed much except to grow older and wider. I, on the other hand, had shaved my long beard off and cropped my hair to near standard military. He didn't know that I was the guy in the beat-up Land Cruiser from years ago that he had met on the roadside, but I knew who he was. I watched his inflated ego flourish through the first day of the two-day class and then put an end to it very quickly.

While lecturing to the class I mentioned that I often tracked while driving and began to relate a case of a missing person on an unidentified interstate. As I began to work through the early parts of the story to illustrate the Eye of the Tracker and Accidental Tracking Cases, I used the word "amiss" several times in place of the word "track." While doing so I looked directly at him, and at that moment time must have frozen in his world. He sank in his seat as the look of horror washed over him, much like I had seen when I first tapped him on the shoulder when he stood by my truck many years earlier. But I kept the story very vague so that only he and I would know the truth. After all, as long as he knew that I knew was what was important to me. It was funny to watch his whole personality change on the second day of the class. As if by magic, the word "amiss" suddenly meant a world of difference to both of us.

THE CONTINUITY

No, I do not go out accidentally looking for dead bodies or illegal graves, but more than a few times they have found me. Certainly I am frequently called in by law enforcement to find an illegal grave; after all, it is part of what a tracker does. An illegal grave is nothing more than a huge footprint. Fresh illegal graves or very old illegal graves are the easiest to find because they mark two sharp contrasts in the overall landscape. A fresh illegal grave, unless the gravedigger is very good, shows some sort of natural hump in the continuity of the landscape, while a very old one shows a depression. This movement of earth from a hump form to a depression is from the natural rotting of the body. It is the "continuity graves," as I call them, that prove the most difficult to define. Yet with a trained eye anyone can begin to pick up the continuity grave sites. A continuity grave is when the decay process causes the earth to assume a flat surface on the land, equal to everything else around it.

This process of the ground going from a mound of earth to a depression can take as little as a few months or up to several decades, depending on the conditions of the soil. The overall decomposition process is affected by type of soil, type of microbes, heat, cold, dampness, and so many other factors that learning to pinpoint an exact time frame is an extreme process. In fact, I've seen graves located less than a mile apart, made at the same time and virtually in the same type of topography, with vegetation decay at far different rates. I found in later soil analysis that the one grave was located in a slightly more acidic soil than the other, yet both were in the acidic sandy soils of the Pine Barrens. In extremes, such as peat bogs, some bodies do not decay at all but mummify, leaving very little depression if any.

I am often asked by law-enforcement agencies how I learned so much about illegal graves, and my answer is always the same, "by doing." I love the horrified looks on their faces when they hear that response. Actually the learning process is quite simple. For years Grandfather had Rick and I do the same thing that I do for my law-enforcement classes today. Months and sometimes even years before a law-enforcement class begins, my instructors and I will collect fresh road-killed deer and various other animals from the road department. We will shave some, leave some with the hair on, and even dress some of the deer in human clothing. We will then bury the animals at different locations, depths, and times, all carefully recorded in journals. During the time frame before the law-enforcement class begins we will take frequent pictures of the grave site. Armed with the photographs during the class we can show how the ground has gone from the mound to depression condition and in what period of time. Generally these photos are taken once a week; sometimes, in fast-decaying soils, the photos are taken once a day.

During a class we will have several grave sites that are each meticulously processed for all forensic evidence and identification. This way the officers that participate in the class will know many of the shapes and forms that grave sites can take during the decaying process. They also learn how to properly remove the remains from the ground while preserving critical evidence and sometimes even footprints of the gravediggers. In the early classes it is also critical that the officers learn to identify the natural lay of the land so that they have a comparison to unnatural land formations. I will also go on to mark old outhouse holes, compost holes, and other man-made depressions or mounds so that the men will know the difference between these decay areas and those of actual graves. We will also teach the psychology of burial, where the officers learn why criminals select certain locations for graves, whether for convenience, time, or desperate acts.

This grave-learning process is further defined by using various types of tools. Graves are dug using everything from the bare hands to sharp rocks, sticks, shovels, rakes, and all manner of other possible digging devices. Each digging device leaves its own unique track and damages the surrounding landscape in its own unique way. After all, a grave is not an island unto itself. There is both the way into the site and a possible second way out. There is the damage done to the surrounding vegetation and landscape, both by the digging tools and the dragging of the body. Every level is taken into account. Many of the shallower grave sites where decay and gasses leak from the soil will show a marked increase in scavenger tracks and sign. We even show how the vegetation growing over the older graves are lusher than the surrounding vegetation, being well nourished by the nutrients below. This effect on the vegetation can also have the opposite effect because of too many nutrients, especially growing over fresh grave sites.

For all of my classes it is always the continuity graves that persistently give them the most trouble. Here the evidence is very scant and one must look hard to locate these graves. Going partly on instinct, the gravedigger psychology, and tracks or sign, the student learns to eventually identify these continuity graves, but it is never an easy process. So often, all of the signs and tracks are in place, but when one digs beneath the debris and subsoil there is no discoloration or contamination. Discoloration of the soil is defined by the improper mix of various soils. Generally soils are in layers or strata, but when a grave is dug and eventually filled in the top layer of the disturbed area does not match the color of the surrounding undisturbed soil. Contaminated soil is that which is unnaturally mixed with debris that should normally be located on the extreme surface of the ground and not imbedded deep within. The whole process of finding these continuity graves can be very long and arduous.

However, sometimes with the finding of continuity graves I just get lucky or unlucky, depending on how one looks at the situation. After years of locating illegal graves for police, by far the fastest I have ever found a continuity grave was during an accidental tracking case. Just a month earlier I had looked endlessly for a grave that should have been an easy depression to spot, given the suspect accused of the murder, the rapid decaying ability of the soil, and all the other factors, but I couldn't find it. I had searched for days with not even the slightest trace of a remotely possible grave site. Several times after the search had been officially called off, I still went back to the area to search and search some more. I was running out of time and had to go home, and home was a very long journey away. I had decided to stay there for the month when originally called out on the case, because the area afforded great tracking and challenging survival. When I first arrived there I thought I would easily locate the grave within a day or two and then take the rest of the month to play.

I stretched my time to the breaking point, knowing that I would not be able to be on time for a friend's wedding unless I got going that very last day. Instead of leaving directly from my camp and heading home I decided to take one last visit to the site. I pulled my old Land Cruiser up to the place I had always parked and went out to look again, promising myself that I would only spend an hour. After two hours I finally gave up and headed back to my car, very troubled that I couldn't be of any help. Finally, doing what I always do before heading off on a long trip, I gave my truck a good looking over. As I bent down to look underneath at the oil pan a small drip of oil caught my eye as I followed its fall to the ground. I gasped, for as soon as the oil hit the ground I saw it, that small yet definite sign of faintly discolored soil. I was parked on the damned grave. In fact I had parked on that damn grave every day of the search, which was sixteen out of the last twenty-nine days.

Needless to say I spent the next week there with the authorities digging up the remains. I missed my friend's wedding and jeopardized our friendship, but at least the case had been solved. Given the location of the grave it just might never have been found at all. I always take great care in considering all possibilities when entering a crime scene area, especially when I have a vehicle, because it is so easy to contaminate a crime scene. I had chosen my parking place very carefully, but I broke several very costly rules. I assumed that the area I had parked in had already been contaminated since I was assigned the area by the police. By assuming I did not check the ground I was parking on, which was a rule I always followed, especially when parking in the bush. I also assumed that the accused could not have possibly been in that area, or that anyone could be that good to cover a grave that well so close to a well-used trail. I learned a very hard lesson: A Tracker cannot assume anything.

FORENSIC PHOTOS

The Eye of the Tracker is with me all the time. It is not only for things found in the purity of wilderness or in the towns and cities, but also in photographs and even in movies or on TV screens. Inside a building or outside in the woods, seeing through the Eye of the Tracker and asking the Sacred Question is a vital part of a Tracker's daily life. As a person learns to become a Tracker, they see all aspects of life differently. Everything is studied, and unlike our ancient Trackers we have a vast new world to track within. It is the world of the photographs, magazines, movies, and TV. Trackers from my school constantly call or e-mail each other about tracks they saw in the media. After all, the lens of a camera picks up not only the primary subject but also in many cases images of the ground. A great game my students play is to see how many takes are done in any given scene of a movie.

Recently I had the opportunity to work for several months on a major motion picture called *The Hunted,* which was directed by my friend Bill Friedkin and stared Tommy Lee Jones and Benecio Del Toro. Though I was the technical advisor for the movie I was far more involved than most others in that position would normally be. Not only did I aid in the technical aspects of the story, help create the props, the set decorating, wardrobe and makeup, but the characters' skills were based on my skills. Even the knives used in the movie were made by me and the instructors at my Tracker School. Tommy Lee Jones plays the part of a Tracker. The great thing about Tommy Lee was that he was already a good Tracker when I met him so there was very little training needed.

Working with these folks so closely, some of my skills were bound to rub off. Many of the crew showed a keen interest in the tracks I was setting down for the actors to follow. They were becoming aware of how I was handling any ground or brush that would be seen in the movie, knowing that to me any stray track would be easily picked up by one of my students. Every day with precision I made sure that the ground was perfect and no stray tracks were to be found. The only exception, of course, was the tracks I purposely left for my students to find throughout the movie. The crew's tracking ability took me by surprise during a private showing of *The Mummy Returns*. We all had been invited to the showing by Jim Jacks, who was not only the producer of *The Mummy Returns* but also on our movie.

The reaction of the crew to one of the last scenes of the movie was both moving and hysterical for me. I clearly saw what my friends had learned on our set, intentionally or not, by the way they laughed at that scene. Just as the balloon is flying by a sand dune we see an overhead shot of one of the primary actors on horseback. It was only on for a moment but the crew just burst out laughing. Those that I had worked the closest with were nearly in tears. For a brief moment, atop this pristine sand dune, is the lone horseman with hundreds of stray horses' tracks surrounding him. The prop master of our movie and one of my closest friends on the set, Barry, elbowed me and said, "How many takes was that?" Scotty, another close friend, turned to me and asked, "How'd you like to freshen that set?"

As Debbie and I walked out of the movie with Barry and Scotty they couldn't believe how many stray tracks they had seen during the movie. I was very moved that the crew had actually paid attention to the tracks at all. I know that many of the people from *The Hunted* crew will pay close attention from now on to other movies and stray tracks on sets they will be working on. It was obvious how easily people can become captivated by tracking.

Unfortunately to a Tracker, we also see photos of tracks that we don't want to see, especially when having to read the forensic photos of a cruel murder scene. Reading these photos, whether from old cases or from recent killings, is another service I provide for the police and FBI. I also train my military teams to read photos sent from spy satellites, actually enabling them to read the tracks of vehicles or troop movements.

Not long ago I was visiting some detective friends of mine several states south of New Jersey. I was just passing through and decided to pay them a quick visit. Years ago we had worked together on the case of a little girl who had gotten lost and drowned. I hadn't seen them in years and it had been a great but all too short visit. While leaving the building I had to pass by one of the detective's desks and noticed the photos of a dead police officer sprawled on the ground. The forensic tracks all around him were very clear and I asked them in a rather matter-of-fact way if they had caught the murderer. The two detectives looked at me dumfounded. "Murdered?" they asked. They looked even more shocked when I could clearly point out to them the tracks of the killer. In something short of a whisper one of the detectives told me that it had been classified as a suicide. Suddenly my short visit turned into a grueling two days of poring over crime scene photographs.

It was obvious that this was not a case of suicide, which greatly relieved the other officers. I could clearly see where the dead officer had pulled his gun and leveled it to take aim at a fleeing suspect. Unknown to the officer another man had come up behind him and shot him in the side of the head. The way the officer's gun landed beneath him and the location of the wound brought about the death-by-suicide theory. However, the tracks told a far different story. Yes, the dead officer's gun had been fired and there were powder burns found on his hands, but it was the way the gun had fallen that made the wound look self-inflicted. I showed the detectives where the gun skidded beneath the officer and how it came to rest where it did. We also found the photos of the officer's patrol car along a dirt path and the signs of a struggle between him and another man. I clearly could identify the height and weight of the other man, including much more incriminating evidence.

Other forensic specialists were brought into the case, confirming what should have been known long ago. For example, when reexamined carefully, blood-splatter analysis did not fit the suicide theory.

We pieced together the case and reopened its files. The slain officer's family could finally rest in the knowledge that this was not a suicide but a murder. There was far too much evidence in the tracks of the forensic photographs to prove differently, but now the killers had to be found. Even though the case was quite old and precise measurements of any tracks had not been taken, I could use the various elements found in the photos that would enable me to give them a more accurate description of the two men. Certainly the man running away did not kill the officer, but he may know the man who did. Unfortunately my time was up and I had to return and teach a class, but I had given them all that I learned from the photos.

Weeks passed without my hearing anything about the slain officer. Two of the detectives came to the Tracker School armed with several more photographs that they had come across. These new photos proved valuable for two reasons. Not only did they enable me to give the police more information than they already had, but I was able to get the help of all of my instructors in the final reading process. By adding more sets of highly trained Tracker eyes to the photographic study, more detail was brought out and exacting scenarios of movement detailed. I felt proud because while I told my instructors nothing of my analysis, they came to the same conclusions as I, just as easily and almost as quickly. The years of training had paid off and all of their personal study time of photographs finally could be put to use.

The payoff came after months of waiting. Two men were arrested and finally confessed to the slaying of the officer. The detectives confirmed what I had told them of the crime scene and all that had taken place. The information we gave them had almost immediately put them on the trail of these two killers, and it just took a little more time to gather enough forensic evidence to make a case. Whenever a case is unsolved there will still remain the original photos of the crime scene, and within those pictures will be the tracks of the killers and other criminals just waiting to be discovered. Time cannot alter what has been photographed, so the tracks will forever remain fresh and visible to the Eye of the Tracker.

THE ART OF PICK TRACKING

To a Tracker tracks are found everywhere. Certainly some have been rendered unreadable or erased by time or other influences, but most any surface can contain a track, if you know how and where to look. Sometimes, as when dealing with forest litter, a Tracker uses the art of pick tracking, which gives the Tracker the ability to read back through vast periods of time. These tracks are found deep in the debris of the forests or overgrown fields where tracks from years past are still visible. Think of a forest floor as a never-ending collection of decaying forest litter. Leaves fall every year along with bark, twigs, buds, scales, insects, and all manner of debris. Leaves fall and cover the debris of the year before, thus protecting the lower layers from the harsh exposure of the elements. Each season past has become a layer that eventually degrades and rots to become the soils of life. Thus the debris is like the lower soils, turning slowly into lines of time or strata. Each layer of debris strata has its own season, which can be identified through the amount of decay.

It is not uncommon for me to revisit a crime scene area, sometimes months or years after a crime has happened, and use the art of pick tracking to resurrect the old tracks. During a long trip I had the time to revisit an old crime scene that had remained unsolved by the police for many years. I had been on that case but couldn't help them very much. Yes, I did find the killer's tracks, but there was no suspect and no weapon was recovered from the crime scene area. Even though the victim had deep knife wounds we could not locate the murder weapon in the area and figured that the killer had carried the knife out with him. Since I was already in the area I had gone back to the crime scene more out of respect for the victim and to pray. As I sat on the ground by the side of a pond and looked out upon this beautiful forest I couldn't believe it had been defiled by such a horrible murder.

As I was about to get up to leave, my hand felt something deep beneath the leaves. It was a depression, that lay hidden beneath the upper levels of fallen vegetation. Out of sheer curiosity I began to use the pick tracking method to slowly and delicately pull away the layers of debris, being even more careful as I inched my way down. I counted each layer of debris, with each layer going back season after season, until the track finally emerged. My heart pounded because not only was this track found in the same season as the crime, but it showed a vague similarity to the boots worn by the killer. I never forget footprints because footprints are like faces, and this one definitely was familiar. It took me nearly two hours to thoroughly pick through the track until it finally came back into full view. It was in an area that had been searched very little so it had remained undisturbed after all these years.

I began to read through what the pressure releases of the track were telling me. Though they were very light and degraded I could still read much from them. After a few minutes of close scrutiny I could read all that the track was telling me, and what it was telling me made my heart race. It was a print left by the killer at a point where he had thrown something hard and far. I could read the line of travel of the throw and it led right to the little pond that was near where the body was found. Without hesitation I followed the line of travel, seeing the knife in my mind spinning through the air and landing in the pond. Even though it was mid-fall I entered the water and crossed the little pond to nearly the other side. I groped around in the shallow waters until my hand finally came to rest on a little knob. At first I thought it was the end of a stick covered in moss, but as soon as I let my hand slip farther down it I realized it was the knife.

Grabbing a small stick from the bank I slipped the stick through the ring on this bayonet-style knife and pulled it from the water. I knew that there would be little chance of any forensic evidence left after all these years, but still I followed protocol. I carried the knife back to my truck and placed it carefully in a plastic evidence bag. There was a certain satisfaction in knowing that I had finally found the murder weapon after all these years. It had eluded me for so long and now it was bagged, all because of the chance feel of a deeply hidden depression in the debris of time. I know that when I turned the weapon over to the police they too felt the same satisfaction for another piece of evidence in a still-unsolved crime.

ALWAYS A TRACKER

The Eye of the Tracker is always with me. It is a way of life, a philosophy, and a way of looking at the world in a grand new way. It opens a universe to me that would otherwise remain invisible. It solves mysteries and answers the Sacred Questions of life and living. Yes, along with all of the beauty it also reveals all of the ugliness, but I learned long ago that there is still a certain beauty in all that is ugly and defiled. It is the Eye of the Tracker that can take that ugliness and turn it to beauty, for if we can see the problem in the first place then we might find the solution. The wisdom is to always be a Tracker, to look deep into the mysteries of life, to find solutions to all that we find to be ugly or defiled. Grandfather so often said, "There is no failure, provided you learn from it," and I believe that even ugliness can be made whole if we can only learn its lessons.

COUGAR CANYON

THERE IS A canyon in Colorado that wrenches my very soul every time I return to that place, whether in reality or from the silence of my deepest memory. The canyon itself is stunningly beautiful. Its clear cold waters wash down rugged ravines and steep-sided rocky buttresses, dancing over boulders in a splendid array of white-water cascades, pausing in deep clear pools framed by miles of rainbow-painted forests. This canyon, though rugged, is by no means inaccessible. It is a favorite hiking area for many locals as well as seasoned travelers. It affords all manner of possibilities for the adventurer, from simple strolls along its steep scenic paths to extended day hikes. It is a gateway into the vast wilderness areas of the Rockies. This canyon touches the wildness and adventurer in everyone, from seasoned wilderness explorers to innocent children. It is specifically this appeal to children that makes this canyon such a personal nightmare for me.

I often wonder what people think of the police reports that they read in local newspapers or see on TV news programs. Those stories all seem so clinical, so sketchy, never really touching the reality and horror of the actual case files as they were lived. I know that the files on this canyon case and the little lost boy must seem that way to most readers, but there is a dimension that goes well beyond what any report could ever convey. It is the dimension that moves from the "just the facts" aspect of the case to the visceral reality of the Tracker. There is a deeper level of a Tracker's Point of View. It is a way of looking, seeing, sensing, and feeling far beyond what is normal, and this Tracker Point of View cannot be easily taught or learned.

The basic act of gathering evidence can be easily learned, but not at the level that Grandfather would require. As a Tracker, Grandfather demanded a deeper, wider, and more intense search for evidence. Grandfather could see in a glance what would take a dozen skilled investigators weeks to find. Awareness to Grandfather was the driving force behind not only all of his physical skills but the doorway to all that is spiritual. It is difficult to define what Grandfather demanded of me as a Tracker. Often he did not teach a skill directly but lived it and taught it through demonstration. I remember vividly one such lesson in Tracking and Awareness, a lesson I call "Grandfather and I took a walk." It is this lesson that a Tracker must learn to live up to, a lesson that a Tracker must become.

I was sitting by our little campfire with Grandfather when he asked me if I wanted to take a walk with him to the dogbane patch. He needed to make a new bow string and the dogbane would be the ideal choice of cordage fibers. Without hesitation I agreed to go with Grandfather. I knew that some of the best lessons came when we took walks together and I would never let an opportunity such as this slip away. Even though I was only ten years old at the time, every moment I could spend with Grandfather was a gift. At this point I had only known him for three years, but I never grew tired of his company or lessons. This same attitude would stay with me for the nearly eleven years that I knew him, intensifying with each passing year.

We walked the entire way to the dogbane patch without exchanging a word. This did not surprise me; often we would walk for hours and never say anything to each other. Yet even during these times of intense silence I still learned more than I could ever convey in words. By just watching him move, observing his actions and reactions, I could learn volumes, and this short journey to the dogbane patch was no different. There was also an unspoken tension I carried in me at these times, for I never knew when Grandfather was going to ask me a question. I most feared his questions about the significance of things we had just passed.

The journey to the dogbane patch and back came without any questions, at least until we sat back down at the fire pit at camp. Grandfather turned to me and with a sly smile asked, "So,

Grandson, what did you see?" Well, to tell the truth, I was ready for him. I felt so sure and full of myself that I burst forth with information. On my walk I had taken great pains to try and observe everything, being aware of every detail and nuance, just in case he asked me any questions. After all, I knew him very well, all too well, and I expected that question. I also wanted to impress him with all that I had seen and experienced. I wanted to show him that it was more than just a walk, an adventure, packed full of little secrets only I had observed.

Without hesitation I let loose, telling Grandfather about all the marvelous things I had seen. I gloated about all those things I had witnessed, knowing that surely I would impress him with my level of awareness and sensitivity. Looking back I think that it must have taken me a full half hour to explain everything I had observed, which was great for me because the whole walk had only taken us about twenty minutes. I discussed tracks of deer, of mice, of ants, and all manner of birds, snakes, and countless other animals. I talked of trails, runs, bedding areas, feeding areas, and various other animal sign. I also described the vegetation we had passed along the way, not only giving the names of the various trees and plants but also their uses as foods, medicinals, or for utilitarian survival needs. I was so proud of myself that I gloated and with that blaringly obvious gloating my confidence and arrogance built to levels never achieved before.

Finally when I was fully finished with my dissertation and bloated with confidence, I looked at Grandfather and asked him a question I still regret asking to this very day. With a sort of mocking tone in my voice I asked, "So, Grandfather, what did you see?" I should have known better than to have asked anything at all. In my own blinding arrogance I had failed to notice that Grandfather was not very impressed with my description of what I had observed on our little walk. In fact, he looked a little disappointed that I had missed so much. Yet the damage was done, I had asked the question, and asked it arrogantly and mockingly, as if to say "let me see if you can add to what I've seen." In fact, as soon as that damn question left my lips I knew that I had made a grave mistake.

Grandfather wasn't angry. He was amused. I vaguely remember two things: the time of day, which I estimated to be midmorning, and the first few words out of Grandfather's mouth. He began by saying, "Well, Grandson," and began his story about our walk together. It became obvious to me after the first hour that the man saw more in the first five steps of our walk than I would have seen in a month of deliberate searching. By noon I began to wonder if in fact we had actually taken the same walk together, and by midafternoon I was sure that we were never on that same twenty-minute hike to the dogbane patch. By what would have been dinnertime his conversation had finally reached the dogbane patch, and by the time I fell asleep sitting up by the fire he hadn't even hit the halfway point back in our journey. Needless to say, I never asked him such a question again. In my mind and heart, Grandfather's level of intense awareness is something I will always strive for, a vision of awareness that I may never reach.

Yes, I should have known better than to ask such a question of Grandfather. I had seen his capabilities, his ability to read the landscape like an open book, to understand people and notice the little quirks that even their own families would overlook. Many times we would start out on a simple walk to gather firewood or take a swim, only to end up with me sitting captivated for hours listening to him recount all that he had seen on these little adventures. What he saw in a quick glance would always fascinate me, stun me, and subsequently make me hunger for the higher levels of observation and awareness that Grandfather possessed. To me, Sherlock Holmes could not even come close to Grandfather's ability to solve mysteries. After all, Tracking is the ability to observe, follow clues, and solve mysteries, with awareness being the foundation of it all.

I realized from the start that Grandfather never separated tracking from awareness; to him they were one and the same. One could not exist without the other. In fact, Grandfather believed so firmly that tracking and awareness were the same that he had only one word that he would use to describe both; that word was "oneness." At the time I didn't realize that word also included the worlds of the spirit. So to Grandfather, Tracking is something more than following tracks, reading the past, and seeing that animal or human coming back to life in the tracks. It is more than the pressure releases that define even the finest movements of that animal or human. To Grandfather a Tracker even transcends the normal limitations of observations and awareness, subsequently entering more profound spiritual realities. This definition, this vision, of being a Tracker then defines a different point of view, a Tracker Point of View that takes the commitment of a lifetime to obtain.

It is this canyon that makes the Tracker Point of View so crystal clear, so real, and so deep. However, it also brings about the ancient struggle of the Tracker, the struggle to balance cold reality and human emotion. Because Grandfather defined a Tracker as something beyond tracking and awareness and brought in the reality of the spirit, emotion begins to play an important role as well as to create an intense struggle. The canyon further defines both—the Tracker and the awareness, the emotion and the spirit, and the balance that must be struck between those worlds, those realities. For me, the best way to fully understand the teachings of the canyon with all the pain and all the enlightenment is to look back upon it through many eyes, many Points of View. It is only then that I can fully understand and appreciate all that I learned there, all that we learned there and will continue to learn.

To view this canyon through the eyes of a police report would be a lesson in futility. I've seen that report and read it carefully. It certainly is a work of art, as all reports must hold up in court and fit the legal criteria of evidence and facts, but it has no soul, no real substance. It contains sterile words and phrases that could never bring the reader into the greater levels of understanding that Grandfather would have demanded. Like most reports, this leaves the reader somehow removed from the whole nightmare, the raw emotion, and from really knowing what happened there. It leaves out all emotion and feeling, dealing instead with the who, what, when, where, why, how, etc., of the situation. Certainly a report written in any other way would lose its objectivity, but in that form the reader or the listener will never fully understand, remaining unaffected, removed, as if something like that happens to others and not them.

Tracking or being a Tracker, as Grandfather defined it, is not something that can be turned on and off. It does not begin and end every time we are called out for a Tracking case or some criminal investigation. In fact, Tracking becomes so much a part of life that the question, "When do you practice Tracking?" should be changed to "When are you not Tracking?" The answer, of course, would be "never," for a Tracker is never not Tracking. Life itself then is viewed differently by a Tracker. The Tracker Point of View is forever part of our makeup, our personality, and our Vision. We tend to easily see intricate detail; things moving in and out of context and just everyday life in general are observed in a vastly different way than most people would even dream of understanding. We tend to observe people and events very closely, watching every detail and nuance, every action and reaction of people in general. We constantly ask ourselves the Sacred Question: What happened here or what is this teaching me?

The Sacred Question becomes a major driving force in a Tracker's life. We want to know. We need to know, and we accomplish this by keen observation and constant awareness both in and out of the woods. Grandfather would point to a cut limb and ask, "How was this cut? What was it cut by? Was the limb alive or dead when cut? Was the person who cut the limb right- or left-handed, how tall, how strong, how skilled, and how long ago was it cut?" He would pick up a cigarette butt and ask, "Was it put out by a left- or right-handed person? Were they male or female, how old, how strong? There were countless questions about countless everyday things that drove us to understand and live that philosophy of the Sacred Question. The Sacred Question became a hunger within us, a hunger that could never be satisfied, because questions led to more questions. The answers to these Sacred Questions could then be found in only two ways: experimentation and observation.

In any given situation, on a Tracking case or not, the Tracker will observe more and be more aware than even the most skilled modern forensic observer. To most detectives it is enough to find a footprint, measure it, determine its tread pattern, and estimate the time it was made. To a Tracker this is only the beginning. For a Tracker further defines the height, weight, dominance, gender, and strength of the maker. The Tracker looks deeper into the print, knowing the emotional state of the maker, his hunger or thirst, the nuances of the smallest movement, the rapidness of breathing or heart rate, and sometimes, oftentimes, even the very thoughts and emotions that run through the maker's mind and heart. Even knowing all of this is just a beginning, for a Tracker learns to read the greater concentric rings of disturbances thrown by the maker of the track. How that person affected the landscape, the animals, the flow and patterns of life in general. To a Tracker, even the smallest clue, the tiniest nuance, speaks volumes.

Being a Tracker and living the Sacred Question then carries

further into life. We read reports, listen to the news, and watch life in a different manner and more carefully than even the most highly trained investigators. We learn to read between the lines, beyond the sterility of the language and into the soul of the situation. We are fence walkers, able to see both sides of an issue and stand balanced on middle ground. Thus the canyon now stands in my mind as the place of the Tracker's Point of View, the embodiment of the Sacred Question, and the ability to walk that fine line between the rational and the emotional. As I said earlier, to read the police report would be a lesson in futility and teach us nothing of the canyon, the lost child, or the vast depth of the lessons learned from the Tracker Point of View. Instead, it would be best to begin with three of the five Trackers involved in the case. These three, their reports, their emotions, and their Visions will then lead us to the wisdom of this canyon. In these three we will find that balance that a Tracker needs to so desperately find. We begin then with the "official" Tracker report as filed, and then move to the reality of the soul of the Tracker.

TRACKER REPORT #671
Filed October 14
Filed by Tracking Officer Kevin Reeve (preliminary report)
Jerry A. Missing Person Case
Brief Summary before Tracker Team Involvement
On October 2, Jerry A. was reported missing from the South Trail, approximately forty-five miles up Cougar Canyon off of State Route #14. Jerry is three years of age, weight approximately

thirty-five pounds, and is thirty-six inches tall. He has dark hair and dark eyes. He was last seen in a blue hooded shirt with a brown fleece vest and sweatpants.

A group of eleven hikers, including Jerry, planned to hike the South Trail and arrive around 1200 hours at the trailhead. Jerry was hiking with a group of family and friends when he became separated from the group. The group was strung out along the trail, and had fractured into a slow group and a faster group. Jerry was with the slower group and ran ahead to catch the faster group. He never caught up with the faster group. He was last seen by two fishermen around 1245 hours. Searchers began searching within forty-five minutes of his leaving the slow group. An organized search began within two hours. For the next several days, an extensive search was conducted involving local trackers, dog handler teams, general searchers, and divers. No sign of Jerry was discovered, not even a track beyond the point last seen. The search was finally called off on Saturday, October 9.

TRACKER TEAM INVOLVEMENT

We offered our services to the sheriff's office on October 6 and were told that trackers were already on the case and they would contact us if we were needed. When Tracker Inc. did not hear back from the sheriff after several days, we contacted Albert A., the boy's father and custodial parent, on Saturday, October 9. We offered our Tracking services at no cost to him or the county, which is the policy of Tracker Inc. He encouraged us to come out and see what we could add to the resolution of the case.

Members of the Tracker Team:

Members of Tracker Team Alpha are made up of police, FBI, military, and public instructors from the Tracker School, a school founded by Tom Brown, Jr., in 1978.

ALPHA TEAM MEMBERS INCLUDE:

KEVIN REEVE, director of the Tracker School

DAN STANCHFIELD, instructor at the Tracker School

NATE KEMPTON, veteran student and Tracker from the Tracker School

DEBBIE BROWN, president of the Tracker School and founder of the Tracker Search & Rescue Teams

TOM BROWN

NOTICE TO ALL LAW-ENFORCEMENT AGENCIES

The bylaws of the Tracker Search & Rescue Teams forbid any form of payment or gifts for their services. Furthermore, the members of the Tracker Search & Rescue Teams wish to remain anonymous and the use of the Tracker Search & Rescue Teams and its members' names cannot be used for any media purposes without written consent from the Tracker Search & Rescue Teams' board of directors.

DAY ONE—TUESDAY, OCTOBER 12

Kevin Reeve and Dan Stanchfield arrived at Denver International Airport on Monday the eleventh and were met by Nate Kempton. They arrived at the general base camp early Tuesday morning and met with the sheriff in charge of the search-and-rescue case. The advanced Tracker Team members requested permission to continue the search in Cougar Canyon. The sheriff granted permission and sent the Tracker Team to the Emergency Services Command center to meet with the specialist in charge of the command center and the search efforts. The specialist briefed the advanced Tracker Team and provided us with a letter of permission. Tracker Team then proceeded to the trailhead at the base of Cougar Canyon. There we were met by air force personnel, who guarded access to the trailhead. They were protecting access to the site of an air force helicopter that had crashed during the search earlier in the week. By midafternoon, the advanced Tracker Team began hiking the trail on the east side of the river, undertaking basic reconnoiter and looking for tracks and other evidence off the side of the trail that may have been missed by earlier searchers. The advanced Tracker Team did locate what we believed to be one of Jerry's tracks on the east side of the river before Camp One. Jerry had apparently climbed up on a rock outcropping covered with moss, which was a perfect tracking medium for identification and preservation. This track was approximately one-half mile north (before) the PLS (point last seen), so it was used as a reference track to establish any further track aging.

The advanced Tracker Team returned from the search area and had a very lengthy and detailed phone conversation with Albert A., the missing boy's father. The purpose of the conversation was to establish a profile of the missing person. The profiling process is a lengthy series of questions that establish the boy's preferences, attitudes, abilities, and physical demeanor. A complete physical description is reviewed as well, looking for wear patterns on other shoes, a description of the shoes the boy was wearing, and any other information that may be useful. The complete profile of the boy was obtained and every aspect of the initial advanced Tracking Team research was fulfilled. It is essential that the Tracker Teams know the habits and abilities of the missing person intimately, subsequently giving the Teams insights into decisions that person might make. The advanced Tracking Team is the SOP (standard operating procedure) of all Tracker Search & Rescue Teams so that Tom Brown's time and skill would not be wasted when he begins the tracking case.

DAY TWO—OCTOBER 13

The full Alpha Tracking Search & Rescue Team, now including Tom and Debbie Brown, arrived at the trailhead at 0800 hours and began the intensive search/tracking process. The Team was immediately able to locate two of Jerry's tracks in the parking lot. These tracks were useful in determining track aging and provided the reference for other tracks that we found farther along in our search. The tracks found here matched the shoe size, pattern, and wear marks indicative of Jerry's footprints. All of the "indicator pressure releases" (tracking fingerprints) were found within these tracks, leaving no doubt that they belonged to Jerry.

We searched up the canyon and it immediately became evident that the area had been searched very heavily. There were searcher and dog prints everywhere and these overlaying tracks would have eradicated any of Jerry's tracks. The Tracker Team, determining the extent of the searching and damage to the landscape, decided to follow SOP and get farther outside the heavily searched area. This way the Team could cut track and speed the process of finding a clear set of tracks that had not been trampled by searchers.

The Tracker Team got to a point where there was a small beach on the river, approximately two hundred yards south of the helicopter crash site. Here we discovered another one of Jerry's tracks, which indicated from the pressure releases that he had thrown stones from that place. One of the stones that he had thrown still lay on top of one of the large flat river rocks, which left small skid marks on the dried slime covering of the rock. There were also several other prints that were found, indicating a series of throws.

We moved farther south on the trail when we came to an easy crossing area of the river, which a three-year-old could easily have crossed without even getting his feet wet. We decided to cross the river at this point and cut for tracks and sign, a standard procedure to rule out the possibility of a river crossing. It had been indicated to us that the prior search teams did not think that Jerry could have crossed the river, subsequently the far river bank had not been thoroughly searched. This was confirmed when the Team reached the far bank and found little evidence of any heavy searching. We did find indications of two searchers and one dog team that we aged to be from Monday, October 4. As we further cut in from the river for sign and tracks of Jerry, Tom Brown picked up a set of tracks that matched Jerry's. We tracked here for the rest of the morning, following Jerry's tracks as he wandered along the river's edge on a deer trail and made one probe up the mountain, apparently looking for the trail. It was obvious from the confusion found in Jerry's tracks that he assumed he was on the original side of the river as the path he came from. At this point he did not realize that he was on the wrong side of the river.

Tracking up the hill was particularly problematic because one of the searchers, a female with a size 8½ shoe, also walked the same way up the mountain. From her tracks she was a dog handler and was following the dog as he went uphill. We suspect that the dog may have been mildly interested in the residual scent left behind by Jerry but she did not recognize the interest from her dog. Given the heavy game traffic in the area, the dog would have had a great deal of difficulty scenting much of anything belonging to Jerry. From the age of the tracks it appears that the dog and handler came by on Monday and the earlier tracks belonging to Jerry were left on Saturday, October 2. In fact, all of the tracks we found were aged to Saturday afternoon, the latest being no later than 1600 hours on Saturday, October 2. We found Jerry's tracks several times, however as we were able to recreate his journey, he appeared to be traveling to the north (downstream) for a while, made another probe up the hill, turned and went back down the hill, continued downstream for a while, then finally turned back to the south (upstream). It is at this point that his trail joins a heavily traveled game trail and the heavy game traffic obliterated his trail.

Not having any physical evidence of anything other than Jerry's trail on the west side of the river, and not having any tracks older than 1600 hours on October 2, we searched the river. The river search was designed to cut track, trying to find another crossing point or other tracking evidence outside of the heavily traveled game trail areas where the tracks had been obliterated. We found no sign or track of any kind along this side of the river, either up- or downstream, for more than a mile in either direction.

At 1200 hours we returned to the area on the east side of the stream and tracked the area near the PLS. We found one track on a rock covered with moss approximately fifty feet north of the PLS. We found no other tracks belonging to Jerry on the east side of the river. The area around the PLS was particularly well searched but we needed to check to see if a stream crossing might have been attempted by Jerry here.

We again decided to work the area south of the point last seen on the west (opposite) side of the river. We began cutting for tracks and discovered an abandoned cabin about thirty feet from the shore. We found three clear compression tracks here that matched the size, shape and pressure releases of Jerry's shoes. The stride measurement also matched that of Jerry's at ten inches. It is important to note that one of the dogs also entered the corner of the cabin we believed Jerry entered. Again, the dog ends up on top of Jerry's tracks. The scent may have been extremely faint, but it was enough for him to follow, even though the area was also heavily used by game. At this point, Jerry's tracks headed north (downstream) again. We followed these tracks as they moved just off the shore on a game trail. They eventually joined up with the tracks we found earlier on in the morning. Again, because of the amount of game moving in this area, we were able to pick up only a few tracks or partial tracks belonging to Jerry, every twenty to thirty feet.

We noted a large number of mountain lion tracks in this area. There are two lions with overlapping range in the area. We found tracks of a female with three-by-three-inch pads in front, and a stride of eighteen inches in a normal walk. Given the depth of the track and tracking medium we estimated her weight to be ninety-five pounds. Her primary range appeared to be on the east side of the river and her traffic on the west side was somewhat limited. There was a large male mountain lion that had the west side of the river as its primary range. His tracks were abundant on the west shore in the area where Jerry's tracks were found. His tracks measured four and a half inches by four and a quarter inches. The stride measured twenty-eight inches in a normal walk. This is a big mountain lion by most standards. We estimated the weight between 180 and 200 pounds. We have been told by the locals that this would be a record cat for this region. It is important to note that the range of weight in most field guides for a mountain lion is between 80 and 180 pounds. A large print for a large mountain lion is three and three-quarters to four inches. We did get a visual of the cat as it moved across some of the rocks by the river in the late afternoon. Tom estimated its weight to be close to the 200-pound mark that we had previously estimated. We also gathered scat samples from old scat that measured one and one-fourth inch in diameter. It contained elk and deer hair and bone particles. In addition, we found the remains of an elk, mostly leg bones, the femur of which had been bitten in half and the marrow sucked out. The teeth marks were those of a large mountain lion's canines and molars. We estimated the age of the kill to be three weeks. After searching the area from the cabin north about four hundred yards and uphill about two hundred yards, we began to lose light and hiked out around 1830 hours.

It is important to note that "stalking" tracks of the large male mountain lion appear to be following those tracks belonging to Jerry.

DAY THREE—OCTOBER 14

Tom and Debbie Brown returned to the Tracker School in New Jersey, leaving the advanced Tracker Team to wrap up the tracks and brief the sheriff's office to its findings. The advanced Tracker Team met with the sheriff and his assistant in the morning. We hiked up to the final tracking area to show the sheriff and his assistants our findings, the marked tracks belonging to Jerry, and the mountain lion tracks. After finishing the briefing we returned to the area we had last searched but abandoned because of nightfall, the day before. We had been directed by Tom Brown to research the male mountain lion, looking for trails, other kill sites, dens, and any other evidence of the big cat's behavior patterns. We established three more kill sites, all of varied ages.

There were also three more piles of bones from elk, including bones and hair belonging to a beaver and various small game. A key for us was that in all the sites, there were few bones left intact by the mountain lion. In other words, the lion was consuming nearly all of the animals it was killing. No additional tracks belonging to Jerry were found.

SUMMARY
POSSIBLE SCENARIO, JUSTIFIED BY THE TRACK
EVIDENCE FOUND

Jerry crosses the stream, just beyond the point where he was throwing stones. He climbs up onto the opposite shore not realizing fully that he has indeed crossed the stream. (This disorientation often happens to very young children when they are distracted by play and adventures.) Once on the opposite shore, he hikes around the area, stopping at the corner of the cabin momentarily. He walks farther downstream, makes a stab uphill to find the main trail he seems sure is there somewhere. When he does not find the trail, he continues downstream for a while but then abruptly turns back upstream. From indicators in his tracks here, he is very disoriented and fearful. By now it is late afternoon and the search has begun in earnest on the other side of the stream. However, because of the noise from the stream, he is unable to hear the calls from the searchers. Panic has begun to set in and he sits on a log for a short time, possibly trying to orient himself.

From this point, there are two possible outcomes. The first is that the lion finds him. The lion would likely have consumed him entirely, including clothing. In this scenario all that would be left would be the shoes, and perhaps a part of the skull or possibly small bone fragments and shredded clothing. Given all the activity on the opposite shore from the searchers, the lion may have retired to a more distant location to eat his meal. Although mountain lions generally eat within forty to a hundred yards of a kill site, we have seen cats carry small to medium kills some distance if there is commotion in a feeding area. It is possible that he consumed the kill in one sitting, or if not then easily in two days of feeding. He would have buried the kill overnight and returned the next day to finish.

We suggest this scenario because it best explains why no trace of Jerry has been found. It also explains why the necro dogs were unsuccessful in locating anything. There would have been nothing left behind. We believe that if the lion were to be tracked for some time, and the den area found, additional feeding sites located, and scat samples gathered, it may be possible to confirm this scenario. The clothing would show up in the scat eventually.

The other possible scenario is that Jerry returned to and attempted to cross the river. In this scenario, Jerry's body would eventually show up in the river. So far, despite the many searches by dive teams and our own probe of the stream, the body has not materialized. This makes this scenario seem least likely.

We find no evidence that might suggest an abduction or foul play of any type.

Kevin Reeve

EXCERPT FROM KEVIN REEVE'S JOURNAL

Tracking in this situation was incredibly emotional for me, and the reason was simple: I too have a three-year-old son. My son has the identical sense of wonder in the woods that Jerry exhibited. I had followed and tracked my own son dozens of times as he wandered in the woods ahead of me when we walked in the woods together. I knew exactly what a three-year-old boy does, how he moves, what he sees. I had seen it before so many times. Then I saw the mountain lion tracks. The tracks were huge, some of the largest I had ever seen. This was not good, in fact that dull ache in the pit of my stomach dropped and intensified. We had a pretty good idea now of what we would find at the end of the trail. Or should I say what we knew we wouldn't find—a body. As we continued to search the area, we shifted from tracking the boy to tracking the mountain lion. It was obvious now that the mountain lion was stalking the boy. Tracks don't lie. The time the tracks were laid down were the same, the area was the same, and the mountain lion was clearly stalking along the game trail. I ached to have the tracks tell me something different. I ached to have Tom tell me something, to confirm or deny the tracks, but he remained cold and calculating. I hate him sometimes. He tracks more like a machine, unmoved by the emotion that strangles most of us. If only I could find a clue into what he was thinking or what he was feeling, that is, if he was feeling anything at all. So often he told me that a Tracker must be a fence walker, never giving in to emotion, never giving in until the very end.

I have a lot to learn. Even though I have practiced tracking for countless thousands of hours and have been on many tracking cases, the most difficult thing I have yet to learn is to remain outside of all emotion. Tom has never given me a hint as to how he does it, how he hides that emotion even from himself. But Tom is much like Grandfather, leading by example, giving only subtle nuances of answers that forever demand my rapt attention. To me it is much the same as a battle-hardened veteran, able to turn it on and off at will, but forever tormented by the hidden demons and anguish that remain buried deep within, surfacing only in times of quiet reflection.

Kevin Reeve

EXCERPT FROM DEBBIE BROWN'S JOURNAL

Ten days after Jerry went missing we were on the case. I was thrilled to be on my first tracking case with Tom. Though Tom had sent me out on tracking cases before, this would be the first time we had tracked together. I wanted us to be heroes, to go in, take a few moments and come out of the woods carrying an alive but very hungry Jerry. Children are so close to their instincts that there is hope. Kids don't mind getting dirty and covered up with leaves and debris like adults do. They couldn't care less what they put in their mouths, so it was possible. This was the hope we all took with us.

The ride from the airport was quiet, but as we drove farther and farther away from the cities and the suburbs, the landscape changed, the mountains began to grow from the earth. I remember thinking, God, please don't get any deeper into the wilderness. But the ride took us far from major roads, and there were no homes for many miles. This is when my mind of a Tracker shifted to the mind of a mother. I began to obsess over the thoughts of my own children, who at the time were six and three. My three-year-old, River, was the same age, height and weight as Jerry. I stared blankly out the window of our rented Suburban.

We stopped at a small convenience store and spoke with two people who knew Jerry's father. We tried to gain as much information from him as we could. The normal "what kind of shoes was he wearing" question was asked. The couple had a picture from a department store catalogue of the exact shoe he was wearing when he went missing. They were white sneakers with a picture of Simba from *The Lion King* on the side. At that moment all things stopped, including my heart. Tears formed in my eyes. They were the exact shoe and size that my son, River, has. A deep pain spread through my body and immediately I tried to understand what that would be like for me to lose River. Only the imagination of real pain was what I was feeling. Panic set in. The interview was taking forever. I asked if I could use the phone. I called Nancy, our boy's nanny, and asked her to find River's shoes, put them on the copy machine, make a copy and fax the picture of the soles. After an excruciatingly long ten minutes I had in my hands a fax of the shoe pattern I would be looking for. I counted every line in the soles and burned them into my memory, especially the tip of the shoe, where most three-year-olds wear away their shoes first, until they learn to become heel walkers like most of society.

I couldn't stop my mind from looking at the shoe and seeing River's legs, body, arms, and face connected to those shoes. His eyes were scared, his body trembling with cold, his face gaunt from dehydration. My bottom lip started to vibrate uncontrollably. Suddenly I was thirsty. All of this was taking too long and yet we had to wait at the store to get clearance from the military because of the Huey that had crashed on the trail searching for Jerry. Patience was not a virtue and I was pacing. Get Jerry, find him now, we're running out of time. God, please let him hang on for a few more hours, we're coming.

Clearance was given, and we were back in the Suburban. The conversation was about the behavior of three-year-olds, their psyches, their perceptions, and their habits. I stared out the window as the landscape rose. A few miles up the road we pulled into a small parking lot for hikers and campers. I was amazed to find no one around. There were a few abandoned canvas buildings that were at one time very active but now stood blowing in the wind unmanned. No dogs, no people on horseback, no police, and no herds of neighbors wanting to do anything they could to help. It was all so sad. They had given up.

At the trailhead was one lone guard, and again we waited. He spoke into his walkie-talkie and signaled us ahead so that we could pass the Huey. We signed a clipboard and were told no photographs and take nothing from the scene. Whatever, we weren't here for a report on the military, just let us the hell in to search for Jerry! We started up the trail, which was wide enough for four people to walk together. The trail started to narrow and climb upward. Before the canopy of the trees had blocked my view, but now as I looked up I had never seen such a tall ravine. On either side, a mountain rose up into the sky. They were huge. The tops were straight cliffs down to the river. The trail looked worn and had been used often. Our pace was fast, we had to get to the 'point last seen' because all tracks were obliterated except for the people who passed last and this was the only way in.

About a half mile in came another guard, who also made us sign a clipboard. This clipboard wanted to know the time of day and who gave us permission to enter the scene. The hell with this scene, this clipboard, the time, we're running out of time. We need to find the little boy and don't give a damn about your Huey. We circumvented the trail around the edge of the mountain because the helicopter was lying across what was left of the trail and forest. The treetops had been sheared off and a few trees were down. There was no fire and this huge aircraft had weaved a spot on the ground with very little disturbance to the overall environment. I couldn't take my eyes off the thing because it was so out of place, so mechanical, so surreal. We knew that no one had died in the crash and could see why. The body of the helicopter was intact with a few dents and scrapes, but the rest of the Huey lay all over the place. It had broken apart in every direction. One blade had fallen a hundred yards away; another was still attached but bent in half. We had to break trail to get around it and that pissed me off . . . another piece of time wasted.

We finally got past the downed aircraft and started climbing. The trail narrowed to only wide enough for one person to go at a time, but we still climbed higher. The steeper we got the fewer trees there were because we were now scaling the edge of the cliff. The ground around became unstable, the trail was becoming a rock outcropping. One slip and I would go careening down three hundred feet straight into the river. I got really angry; how could anyone let a three-year-old walk this trail alone? I had to hug the ground in some spots, it was so steep. I imagined my son falling from this spot. It was so high and steep that nothing would survive a fall like this. I'm not afraid of heights but this was life or death. I wanted to make sure each step was secure before committing all of my weight to the next. No one spoke and the gravity of this wild place was dashing our hopes. I began to think about finding the body.

I imagined finding Jerry's tracks, following them and reaching his lifeless body. It played over and over in my mind. It was always the same. Jerry would be facedown and I'd have to turn him over and when I did it was River. I couldn't stop it from replaying. I tried to picture him alive and think there was still hope, but my mind would just go into rewind.

We started the descent and the trail opened. This was a safer place to be and we knew from the yellow ribbons on the trees that Jerry had made it past this part of the trail. I stopped to read one of the ribbons and I wish I hadn't. It said "dog, body" on the ribbon, which means a dog got the scent of something dead. I hung on to the thought that dogs do a great job as long as it hasn't rained and a lot of people haven't gone into the area. It had rained and snowed and there had been hundreds of people searching. Sometimes dogs make mistakes.

We arrived at the ribbon that said "point last seen" and split up. I wanted to be with Tom, the master at tracking, but I knew even he needs help and most of all he needs me to do the job and find the tracks. Quickly surveying the ground for any partial of Jerry's tracks, I am somehow secretly searching for River's tracks. I see him skipping and stopping to look at the river that raged nearby. I see him stopping to look at the bugs on the ground and playing with the dirt. Now I realize that I cannot separate Jerry's tracks from River's. They become one and the same and I began searching all the areas that River would go and what he would do. It was becoming too close for me and I sought out the others. They had formed at the edge of the river. I don't remember who, but someone said, "This may be Jerry's track." We all waited for approval from the master. I pulled out the copy of the faxed track. It was the same size and shape. The track was deeper at the toes. Tom said the track showed that Jerry was throwing rocks into the river at this point. When he said that it all made sense. When you throw something, all of your weight goes to the tip of your shoe to counterbalance the weight being moved forward. I see River throwing rocks from those tracks. I know I am unable to separate my emotions from the track. I still to this day see River in those shoes and on that trail, cold, hungry, and scared. As a mother I wanted to find him, one way or the other.

We stood behind the track looking for the next one and

where it would lead us. It led closer to the river and we all knew this is where he crossed. There were perfectly placed rocks in the water that would allow a three-year-old's legs to jump to each one to the safety on the other side. All four of us went over. Within moments we found another track and another. There were also the tracks of a dog and a construction boot and we knew that someone came over to search the area. We had been told that Jerry would never cross the river, but we found his tracks over there. At this point his tracks told us he had no idea he was lost and that the others were searching for him. Tom went back across the river to the original track for an experiment. We put our backs to the river and he shouted from the other side, "Jerry," as loud as he could, but none of us could hear him. The sounds of the rushing water drowned everything out, even screams from just a dozen yards away. There was no way that Jerry could have heard the searchers calling for him. Not even a whistle would have traveled over the sound of the river.

The next track we found was that of a cougar. It was easy to see. It was huge. No dog would have a paw that big, and anyway cats have retractable claws, as this one did. I remember that we were told that Jerry weighed 34 pounds. River weighed 33 pounds on his last doctor's visit just two weeks before the tracking case. Thirty-four pounds for a big mountain lion is nothing. They can carry a 125-pound deer high up into the rocky ledges of the mountains. I listen to Tom as he walks us through the tracks of the cat. The tracks were days old and they were carefully placed tracks. The cat was stalking Jerry. Tom pissed me off. His voice was so clinical, analytical, and unemotional. He recounted the tracks as if it were nothing more than a mountain lion stalking a rabbit. Mr. Ice, I remember that he told me the police had named him that back when he was in his early twenties. Well, he earned it again.

Now everything shifted to finding the den or kill site. Cats will carry their prey near their home if they can and we split up again. I chose the trail on the riverbank and started heading back toward the trailhead. There were many places to afford safety for a big cat. Huge rocks were everywhere and erosion had removed the earth from underneath them. There were many crevices and places to hide. I started searching all of them. I was scared of finding Jerry's body in them, but now I was scared of crawling into a den with a big cat. I tried to make as much noise as possible, hoping the cat would have two ways in and out and that he would take the alternative route of escape.

I climbed into one, scanning the darkness for a movement and listening for a growl. There were areas that were too dark to see and I had to crawl into them and feel around, half hoping to find nothing and half wanting to find Jerry. None of these hiding places had any evidence of the cat or Jerry and nightfall was coming. Defeated I began to work my way back to the group. The walk was slow, I had nothing. It broke my heart to turn back. I wanted something, some evidence of Jerry to return to his family. I couldn't imagine what his parents were feeling and it hurt to stop searching. If I couldn't find Jerry at least I could help point the direction for the other searchers to follow. All I can tell them is that a mountain lion had stalked

Jerry on the other side of the river. I couldn't give them closure.

I remember the flight home. My eyes fixated out the window, tears flowing. I heard laughter between two passengers and thought, How can people laugh when there is a child out there missing? There is a mother and father out there that have no clue what happened to their three-year-old son. Mr. Ice sits beside me, a blank expression on his face, but something has shifted in his eyes, deep and hidden. He turns to me and answers my question before I even ask. "I've seen too much of this, too many children, too much death and pain. It doesn't get any easier. It only grows more difficult, the pain, and the pain makes us strive to be better Trackers. The lion, or the river, took Jerry to a place we cannot follow, and not being able to follow drives me to find a way."

As I thought about what Tom said, I began to understand what drives him, and what makes him the way he is. I so often accuse him of brooding, never fully knowing the reason why until now. I think back through all of the stories he has told me, all the tracking cases, and all that he has seen in his life. It wasn't until I too searched for Jerry that I finally understood the burden he carries deep within him, buried in the place of the brooding I accuse him of. It surprises me that he is able to keep it hidden, especially during a tracking case, where emotional involvement clouds both the physical and spiritual act of tracking. I too am beginning to brood, brood over all the things I have yet to learn, brood over how I could become a better tracker, and brood over the loss of such a precious little boy.

Debbie Brown

MY JOURNAL

I knew even before I entered the canyon that there was little hope and I grew certain of it the first time I touched Jerry's track. A track is like the end of a string and at the far end a being is moving, and when touched it connects me to the far end of the trail, to the maker of the track, and thus I knew that its maker was gone. It is a struggle to become removed from all emotion, for to track is to become that which you are tracking. To feel the pain, the confusion, the hunger, and the fear of this little boy, etched in every track, only complicates and compounds the struggle for objectivity. There is a place within me, forever hidden, that entombs all emotion when I am tracking. The only time that door is open is when I need to feel the emotion of the being I am tracking, and then I quickly close that door again. It never gets easier to open and close that door. Many times when my guard is let down the door will open, casting me into a place of emotion and pain I don't want to be. This place within me subsequently must always remain guarded, closed, and cut off or it will devour me.

The struggle intensified when I crossed the river and confirmed Jerry's tracks. I could feel his wonder and play shift to confusion and fear. A broken twig where he stumbled and fell, his little handprint in the leaves, and the tears welling up in his eyes all built the emotional picture within me and I became the child. He did not know that he was being stalked, nor did he see the big cat, but something had shifted inside of him. A primal fear welled up in his heart but he could not know why. I could also feel the cat, watching Jerry, muscles tense, waiting, stalking, and then waiting again. I could feel the lion's nervousness as he caught wind and sound of the searchers, his ears perked to the other riverbank as he tried to pick up wind-driven scents, yet nothing would drive him from the stalk.

I could feel my heart pounding into my throat as I looked at Debbie, her eyes searching me for answers. Kevin and the others paused to watch me track, begging in some strange way for answers to their own unspoken questions, their fears. I could not give in to the emotion, or let them see that I was hurting inside, hiding my anguish and pain as I felt the child's fear and helplessness pulsing through every nerve fiber of my mind. I fought back the tears with the usual icy stare that I have perfected over all these years of tracking. In the tracks of this small boy I saw my own son's tracks and it terrified me. At that moment in space and time Jerry became my son and I struggled hard to separate the two. I had to fight it back, I had to get back to the track in a very clear and rational way, removed from all the pain and fear surrounding this child, and as usual the struggle within me became an epic journey.

I vaguely heard their questions, their need for confirmation, and I answered with that clinical tone I use when teaching a class. The images of the tracks now seemed surreal. I was recounting both fact and emotion to my tracking team, yet it is not my emotion but the emotion of this little lost boy. I brought him back to life in those tracks, his every movement, his fear, his thirst, and his confusion. I brought forth the stalking cat, its lethal movements and its total disregard for the distant searchers. So too were the concentric rings etched forever in these tracks; the ebb and flow of birdcalls and flight, the chatter of squirrel, the distant movement of deer and the ever-warbling roar of the river, all combined in that moment of raw emotion and stark reality. I was reluctant to give the Tracker Team any closure because I knew what would soon happen. I didn't want them to live the same nightmares I have to live with, for it's better that they don't know and can only guess. They could then form their own conclusions and their own nightmares and I wouldn't have to be responsible by having them see through my eyes.

I watched as Jerry's little tracks disappeared into eternity, forever becoming covered by the countless animal tracks along the game trail. I knew that at this point to try and follow would be a lesson in futility, a random hit-or-miss situation that might have taken weeks or even forever. The most difficult thing for a tracker to do is to turn from the trail, knowing that all hope is lost. As with so many trails, that turning will haunt us the rest of our lives, filed away into that place that remains deeply hidden within us. I did not have to tell the Tracking Team that it was over. They knew from the tracks, the evidence, their raw emotion and their instincts. I remember my last words to them as we turned away: We are not giving up, only giving in to reality, and that reality is bitter to accept or understand. We walked out in profound silence, each struggling with his or her own thoughts and emotions, enveloped by the symphony of the river and the dance of creation, a eulogy to a little lost boy and the eternal dance of life and death.

DEATH OF A TIGER

It was not the typical way I am called into a tracking case. Evening was fast approaching and I was headed home from an advanced tracking class at our Primitive Camp in the Pine Barrens. It had been a long day of tracking, where students were following fox tracks across the frozen sands of an area we call the Maze. For January it was unseasonably warm, but the thaw had only hit the sunny areas; all that was shadowed had remained frozen. It was a day of challenge for both the students and the instructors. Following these small and light gray foxes of the Pine Barrens, walking across frozen sands, was a tremendous lesson for the students. In reality, it would be like trying to track a human across a concrete sidewalk. The flatness of dust or a slight movement of a single grain of sand defines these tracks.

I was tired from the day of intense scrutiny. Typically I move from tracking team to tracking team, correcting mistakes and giving helpful advice to my students. At this point in their training, even with having completed several prerequisite classes, the average speed was one track every half hour. Eyestrain, especially after this kind of intense tracking, is always a given. Essentially it is like reading a book. After all, the land is like a huge manuscript, a journal, or an open book, written and rewritten every day by animal authors. These trails, tracks, sign, and marks are like letters, words, and sentences of the animal world, just begging to be followed and read. In these pages of the earth there are great stories and mysteries, begging to be deciphered and solved. This had been such an intense day of reading.

Unlike books, however, which are flat and have only two dimensions, tracks are three-dimensional. This kind of tracking eyestrain is unparalleled in the modern world. I am continually reminding my students to take a break from the tracks and trails every hour, sometimes even forcing them to look at the sky or larger landscapes. If these breaks are not taken a series of distortions could set in. These distortions go beyond the simple eyestrain that most readers would encounter. If left unchecked, this strain would lead to a depth-perception problem, where the Tracker could not distinguish depth at all. This then leads to an intense headache and something we Trackers call ground surge. This is where the ground appears to be moving, much like the undulations of the sea. Many times students have become seasick because of the ground surge mirage.

Fortunately, tracking is very much like going to your local gym for a workout. The more one works out, the stronger they become. With tracking, the more you practice tracking the less eyestrain, depth-perception problems, and ground surge you experience. However, like a physical workout routine, if tracking is not practiced routinely then the eyestrain and other maladies return. Yet even with decades of experience tracking every day, no one is immune to these problem. Now at day's end and with helping nearly twenty tracking groups read the trails, I was glad to be driving home for a break. The last thing I wanted to do at that moment was read another track, large or small.

Yet tracking for me is constant. My world is a constant tracking scenario. It makes no difference to me whether I am in the wilderness, suburbs, or city, whether I am walking on a trail or driving in a car, I'm always tracking. As I drove home I tracked, watching for the various trails, runs, beds, pushdowns, and so many other signs that are found along our roadway edges and median strips. Basically I can't stop tracking. The earth is alive with tracks, with mysteries waiting to be unraveled. It makes little difference to me where the tracks are found. The tracks of a pigeon along the gritty fringe of a city street are as interesting to me as the track of a bear found deep in the heart of a pathless wilderness. Tracks, simply, are everywhere.

It was during my drive home from the class that I did something considered strange for me. I turned on the radio. I rarely listen to a radio while driving, for driving is a place I can get lost deep in thought, a type of meditation and valuable alone time. Listening to the radio or music is only a distraction and used only when I want to get out of my thoughts for a while. The only thing I listen to on the radio is the news. As a Tracker it is important that I keep up on current events, especially if a fugitive is on the loose or a person has gone missing. This way I know to expect to be called into a tracking case and make early preparations. The odd thing was that I didn't want to get out of my thoughts during the trip home. In fact, I wanted this quiet time to think about the class. But something deep inside me moved me to turn on the news.

At first I could not believe what I was hearing. Something about a tiger roaming the streets of Jackson Township, which is located on the northern edge of the Pine Barrens. I was stunned. It had to be wrong. Certainly there were tigers in Jackson Township, at Great Adventure Park, but these were safely locked away and escape for them was nearly impossible. Listening on, it was confirmed there was a tiger loose but the Great Adventure people said that all of their tigers were accounted for. This baffled me. Could they be lying? Was it someone's pet that had escaped? I knew that people sometimes kept exotic pets but the notion of a tiger as a pet was ludicrous. I raced home, knowing that I would probably be called into the tracking case to find the tiger.

As I drove home, still listening to the radio, my mind was filled with questions. Yes, questions about the tiger and the concern that it was now full dark, but also concerns about the area the tiger was supposed to be in. My friend Joe Cutter, who heads up the news radio division of 101.5, had said that the tiger was spotted several times near a development. In this report he also said that local Jackson police officers were "tracking" the tiger. This did not surprise me, for we had just finished training three of the officers from Jackson Township during one of my police classes just a month ago. I felt relieved because these officers were good trackers and more than capable of following the big cat. After all, according to the news report, they had tracked the tiger throughout the day with no problem. Possibly I wouldn't be needed at all.

As I pulled in the driveway, Debbie met me and handed me my track pack. The track pack is nothing more than standard equipment that all of my tracking teams carry. It is a small pack that houses a few essentials, especially tools to preserve critical evidence. It is only since I've opened my school that I started carrying anything on a tracking case, but preserving evidence is critical, especially if there is a fugitive involved or a crime has been committed. As soon as I saw Debbie standing there holding the pack, I knew that I had been called in to find the tiger. She didn't allow me to open the door of the truck, but approached the window and passed me the pack. I could tell that she was worried, but I know she would never tell me that she was. After all, she has learned to accept what I do, just as my folks had long ago. In fact I've sent her on several tracking cases when I could not go myself. She knows what needs to be done, despite fears or concerns.

Hardly any words were spoken between us, which is usual for Debbie and me. We have been together for a long time and know each other so well that classic communication isn't always necessary. Being with Debbie is like being with Grandfather; we know what the other is thinking and feeling without the need for any words. In fact, the tracking pack and the tracking team networks are Debbie's creation. She saw a need for skilled tracking teams that I had trained, stationed throughout the country and ready to go at a moment's notice. Everything that a Tracker would need is in the track pack, so all they have to do is grab the pack and get to the tracking case. Many team members also carry a track pack to work or in their vehicle, just in case they are called while away from home. I, on the other hand, have a love-hate relationship with that damn pack.

Debbie filled me in as I sat in the driveway. The big cat was a Bengal tiger, which had escaped from a tiger refuge in Jackson Township. The police would not get close enough to the big cat to tranquilize it and needed my help to stalk close. She also said that the police had requested me for tracking. This really confused me because I had trained the officers and they were some of the best I had trained to date. All the police said was that with nightfall they were afraid of losing the cat. What few people realize is that a huge tiger will not leave much of a track. Their paws are so wide and they walk so stealthily that they leave little trace discernible to the untrained eye. At night the tracking would become extremely difficult, especially given the frozen ground and the type of terrain. Still, I felt that my officers could follow the tiger without any problems.

I was lucky enough to get a police escort most of the way to Jackson. First by the Long Beach Township Police and then by the State Police. Debbie must have called them in, for they were waiting for me at the first turn off my street. Both the Long Beach and State Police had been trained by me many times before and knew the situation. They did not want to see anyone hurt, nor did they want the tiger shot. At this point there had been an order given to shoot to kill the tiger since all efforts to tranquilize the animal had failed. As soon as I heard this news over my police radio I raced against time, raced to save the tiger. At the same time I knew the pressure that everyone was under. This tiger was undoubtedly hungry and posed a very real threat to the community. It had to be found and found soon.

As I hit the outskirts of Jackson Township, I was not surprised to find that my Tracker School and Tracker Team director, Kevin, was arriving also. It was obvious that Debbie and the police had also called him into the case. That is standard operating procedure with our Tracker Teams. Law-enforcement agencies are told to call several of my top Trackers, not just me, when a tracking case evolves. This way if one of us is unavailable, the other will be backup. It was a relief to see Kevin there, even though I had trained the Jackson police several times; I needed someone with his level of skill. After all, he is not only the director of my school, but also one of the finest and most dedicated Trackers I have ever trained.

We jumped from our trucks and headed straight for head-quarters, which was a fatal mistake—fatal for the tiger in the final analysis. Not a word was exchanged between Kevin and me. A major part of my Tracker Team training is concerned with nonverbal communication. Words and even hand signals can become a liability. Not only does sound and motion disrupt the concentric rings and natural flow of the forest, but they also become an extreme liability whenever tracking a fugitive. A Tracker never wants to give away his location, especially to a potentially armed and dangerous fugitive. It is strange for an untrained observer to watch the intimate workings of a Tracker Team in action. Instead of words, the Team uses body language and subtle nuances for all communication. Sometimes, often-times, it appears that we are reading each other's thoughts.

We got to the desk sergeant only to find we were at the wrong place and out of time. We should have been directed to the for-ward communications post a mile away but the message was never forwarded to us. We raced back to the trucks, sidestepping the surge of media reporters, flashbulbs going off and TV cam-eras rolling, and followed our escort out of the driveway. As we turned down the street to the development we heard the shots go off. I felt sick to my stomach, because I knew the worst had hap-pened. The tiger had probably been shot. The detour to the police headquarters had probably cost the tiger its life. I knew instinc-tively that if I had been there just five minutes earlier I could have gotten close enough to the big cat to tranquilize it, but now all hope was vanishing. I prayed that they had missed the shot.

I pulled the truck up to a screeching stop and ran to where a group of officers were standing. Even before Kevin and I got to them I could see the blood on the parking lot surface. Huge claw marks had dug deep into the blacktop at the same point where the bullets had hit the tiger. It was also very obvious, even with only the scant lighting of a lone streetlight, that the tiger had only been wounded and his tracks now led off into the deep woods. Though the line of tracks and the claw marks indicated that the big cat was moving at a high rate of speed, I knew that it had been wounded badly. Now an even bigger problem existed. The badly wounded tiger was now a very lethal threat, and we were on the outskirts of a highly populated subdivision. This tiger had to be tracked and tracked fast.

As we neared the gathering of police officers we were relieved to see two of the officers we had trained. Tony was bent over, feeling the claw marks, just as I had taught him to do, and Bill was sidelighting the tracks with his flashlight, which caused the tracks to flair up with reflection. That moment made me so proud. They were tracking the tiger across solid blacktop, something unheard of by their peers. I knew all along that they could have followed the track right into the night without any problem, but now I was there and involved. All I could do for them now was to track down the tiger and keep it from killing someone. It frightened me to see small children standing on a porch outside a nearby house. The neighborhood had no idea of the lethal ability of a wounded tiger.

I talked with the police for a while and quickly organized a

tracking team. Kevin and I would take the double point position; Tony and Bill, who were heavily armed, would be the wing
position and backup protection; and the rest of the officers
would fan out slightly and become secondary backup. There was
no doubt in my mind that this wounded tiger would turn and
attack if confronted. It would even be capable of circling around
and attacking the team from the flanks or rear. Everyone had to
be at the ready. No shot would be taken if I saw a chance that the
tiger could be taken by a tranquilizer dart, for not one officer
wanted to see that animal shot.

I drove my truck up to the edge of the parking lot to where
the tiger had gone into the thick woods. I turned on all of my
tracking lights so as to illuminate the forest as much as I could.
All of the Tracker Team trucks have an array of tracking lights.
Often we track fugitives or lost people while driving roads or
trails at night. These configurations of high- and low-intensity
lights, positioned at several different angles, make for easy tracking. With the aid of these light positions we can easily pick up
places where people had crossed roadways or trails with ease,
greatly reducing the time needed to find these crossing points.
With the lights now on, at least the first part of our tracking
would be fast and easy.

I stepped off the parking lot and into the thick brush. Blood
was everywhere. The tracks indicated that the tiger was faltering,
bleeding heavily, and slowly dying. I motioned to Kevin to look
down at the track, but he looked at me puzzled. I realized that he
couldn't see it at all and I knew the reason. Like me, he had been

tracking with the class all day, concentrating on the small tracks of foxes on the frozen sands. His eyes and mind were still geared to that size and level of obscurity. Without a word I bent down to the ground and spread out my whole hand and fingers above the track. The expression of shock mingled with realization came over Kevin's face. The track was larger than my entire spread hand, but not very deep at all. Kevin's eyes just stared at me in utter disbelief. Certainly he had tracked large animals before, like grizzly bear and mountain lion, but he had never seen a track of this magnitude before. I on the other hand had tracked two escaped tigers before in Canada and was prepared for their size.

Kevin's expression then changed to one of sadness, for he immediately saw what I saw. The pressure releases of the tiger foretold its slow death and hideous pain. Kevin was breaking a rule by allowing his emotion to overshadow his thinking. I touched his arm and he snapped back into the cold consciousness of a Tracker and the reality of the dangers we now faced. At that one moment in time, I could tell that Kevin had just learned one important lesson of Tracking. He had fought that emotional battle within and set aside his pain. He relearned and finally understood the internal world and struggle of a Tracker, now made manifest by the dying tracks in the frozen ground. It is not something that can be taught, but something that has to be witnessed and struggled with constantly. Though a veteran of many tracking cases, Kevin had to know that pain again and transcend its obscuring grip.

The progress was very slow. It was not because we could not

find the tracks and follow the trail, nor was it because of the thick brush and frozen ground, but because the tiger could be laying in ambush. It could strike from anywhere at any moment. The forest was virtually silent, as if in mourning. Any concentric rings of nature were obliterated by the sounds of crackling police radios, hovering aircraft, and constantly arriving police cars. Without the subtle nuances of nature to guide us, our Tracking Team was in even graver danger than would have been normal. Even as we radioed back to stop all traffic and turn down all radios, we could not read the mood of the forest. Nothing outside of the tracks gave away the position of the tiger. All I could be certain of was that he was slowly dying and in tremendous pain and confusion.

As we moved farther into the forest the lights from my truck began to grow dimmer. Long shadows from the high-intensity lamps and the misty conditions of the lowlands we were now entering cast an eerie glow all around. Only Tony and Bill had worked with us as a tracking team before, the other four officers had not. It was ironic that without any coaching the team moved together, as if guided by our own collective instinct and the night. We moved much like an intense mission in Vietnam, the backup guns ready, and our eyes scanning the horizon, scrutinizing every shadow, and moving with a slow labored caution. Even though the wounded tiger was nearing death, he could still be laying in wait, like a sniper.

Beyond the sanctuary of all the trucks' tracking lights we had to use small red-beam flashlights. Kevin and I took turns feeling the tracks, reading the fingerprints of pressure releases. I estimated the tiger to be close to 430 pounds, large for a young adult male. It struck me heavily that he moved more like a big house cat than a tiger. This world must have seemed strange to him. I could also tell he was very frightened and bewildered. After all, he was born to captivity and never knew the freedom that he now lived. I felt sorry for him, at times almost on the verge of tears, for I didn't want to see him die. The sorrow was often replaced by the anger, nearly rage, for the people responsible for this beautiful animal.

By now we had tracked the tiger for over a mile and I knew it intimately. His actions had no reason now, his life was slowly slipping away with each step, draining from his body in spurts of blood splattered against the brush, ground, and tree trunks. We suddenly had help in our quest that lifted some of the intense pressure from us. A small plane with an infrared sensor aboard was now circling overhead. Our radio confirmed the fact that the tiger was lying down only about one hundred yards from our position. That new and unexpected information was like technological concentric rings, allowing us to move much faster and without great fear of an ambush. According to the aircraft, the big cat was lying off to the side of a small stream.

I felt the tracks knowing now that the cat was only several yards away. For an animal of that size and strength we were not that far away and I grew concerned that if he did charge we might not be able to get a shot off in time. I called for the normal flashlights to be turned on and there in the flood of light, not ten yards ahead, lay the tiger. I could see the faint labored plumes of vapor coming from his mouth, the rise and fall of his chest, and the slight tremors of his body. He could barely lift his head. Suddenly a shot rang out and the bullet cut through the shoulder of the tiger, casting blood, flesh, and fir in a reddish halo about him, and he fell limp.

Despite the protests of everyone, I ran to the tiger. I reached down and felt the faint pulse in his neck, felt the last few shallow breaths and the fire disappear from his eyes. I lifted his great head onto my lap, trying to force back tears. I could feel the strength of his muscles as he twitched in my hands and the last beat of his heart slip through my fingers and into the eternity of night. Kevin was close behind me and he laid his hands on the tiger, riveted in a place somewhere in the emotional landscape of awe and sorrow. He tried to speak, saying, "If only . . ." I didn't allow the rest of his sentence to emerge, for I had been to the place he was now in so many times before. I said, "Time beat us this time and will again. We are a few that stand at the threshold of time and eternity, so revel in it, and learn from it. We witnessed eternity unfold tonight."

Kevin had raced the clock before but had never been so close to slamming the door on eternity. I knew what he would go through, for I lived it and will continue to live it. Now it was his turn. As we walked back to the trucks, the commotion, and the media circus, I could see and feel something shift in Kevin. That unspoken shift, that oneness of time and eternity is a lesson that a Tracker must learn, understand, and live. The tiger had brought him to the brink of time, of eternity, and what it means to be a Tracker. It cannot be taught or learned in any other way, for it must be lived. At that moment of space and time Kevin finally understood me and the driving force behind my soul, for it is the same driving force that now drives him.

The Tracking Team walked from the forest, not as glorious hunters and saviors, but silently and with dignity. None of the crew that awaited us could really understand what we had been through. Our team solemnly awaited the tiger to be dragged from the bowels of the forest. It was so massive that it took seven of us to lift it into the bed of a pickup truck. Its power and beauty stunned us into a state of silent awe and reverence. Its claws, paws, teeth, and fir were so grand, so beautiful. What a loss to the world, I thought. Could there have been a better ending? Could I have beaten eternity? Could I have been called in earlier? Do I have to live with another failure for the remainder of my life? My mind overflowed with so many questions, so much sorrow, and no real answers.

Kevin and I walked for a while with Tony and Bill. I could sense that in their own way they were feeling the same emotions and asking the same questions I was. Upon our return to the truck we saw a group of men laughing, taking pictures of each other with the dead tiger, and examining it with no remorse in their touch. At first I grew angry at this circus, wondering what to say and do. For some reason, without a word and driven by a force or thought somehow greater than myself, I just put my hand on the tiger, looked to the sky and said a soft prayer. Suddenly all the men around me stopped; some took off their hats and bowed their heads, all of them prayed. I don't know what drove me to do that act of prayer and reverence but I know that it touched everyone beyond any words I could have spoken.

When the tiger was finally taken from the parking lot and our initial reports were filed I headed home. The ride home went quickly because I was so wrapped in questioning thoughts. I know I could have stalked close enough to that tiger if I had had enough time, but there must have been a greater reason behind its death. Certainly the men who shot it were only following orders and the safety of the community was at stake. At that point they had no other choice. After all, they had tried to tranquilize it twice and had failed both times. I thought that possibly the reason for the tiger's death was to teach Kevin the lessons of time and eternity, without having to face a dead child as I had had to do, but I felt that this could not have been reason enough to take the tiger's life.

I sat up for most of the night, now shifting my attention and anger to the people that ran the tiger compound. I was enraged that the animal had not been better penned. They should have known that an escape of any sort would only result in the death of a tiger. It would also put an entire community at risk. I wondered how anyone could legally have a tiger sanctuary so close to civilization. From the stories I heard during the briefing session, the tiger facility had been in operation for quite a few years, but the area was now growing and the compound should have been moved. I blamed both the tiger's owners and the government for the death of the tiger. Someone was to blame and it was pissing me off.

An early-morning call shook me awake. The Jackson police needed me to come back. At first I thought that another tiger had escaped but that was not the case. Apparently the owners of the tiger compound were denying that the tiger came from their place and were casting blame on Great Adventure. Great Adventure had proven to the authorities that all of their tigers were accounted for, and they needed me to backtrack the tiger. Possibly, they thought, it might have come from another place altogether. It might have been the pet of a third party. The only way we could tell, outside of DNA analysis, was to backtrack the tiger. Considering that the usual time for DNA analysis would be more than a month, time was a critical concern in this case. Following the tiger to the place of origin would be crucial.

In my mind and heart there is something sad about tracking a known dead animal or human. As stated before, tracks to me are like picking up the end of a string. At the far end of that string a being is living and moving. When the being is dead, it can be felt in a big way. Tracks record every action and reaction, every movement and thought, as if the animal or human has come back to life through the doorway of the tracks. Now the end of the string is death and what once was will never be again. I just could not bring myself to follow those tracks, knowing full well that they were the last hours of such a beautiful animal's life. But it had to be tracked and those responsible either shut down or punished.

I called Kevin and part of my own Tracking Team. Most of the team members were instructors from my school and are among the best I have ever trained. I had no doubt that they could easily track the tiger. The tracking would also give them valuable experience, especially tracking a dead animal. I know that this would become a lesson they would cherish for the rest of their lives. It would be a lesson that could not be learned in any other way. As expected, they all agreed to go and were on their way within ten minutes of my phone call. Knowing that what they found might be needed for a court case, I had them take very valuable forensic evidence as they tracked, including plaster casts, foot maps, measurements, and both photographs and video. This way there would be an important physical record of their results.

The one thing that troubled me, however, was that they seemed to lean toward the tiger preserve as being the guilty party and not Great Adventure or another outside owner. I would not let them go out until they all agreed that they would not believe anything until it was proven by the tracks. They were breaking one of the most important laws of tracking. They believed something to be true before proven by the tracks, thus prejudicing themselves and causing a conflict in sound judgment. I knew from years of experience that a Tracker cannot assume anything, no matter how strong the evidence. A little prejudice will cause costly mistakes. I've seen those kinds of mistakes happen more than I care to remember, not so much with my tracking but with any investigation.

Finally, after feeling confident that they were open to anything the tracks told them I sent them on their way. Because time was no longer a factor and all evidence had to be meticulously gathered and preserved, they spent the entire day tracking the tiger to its source. The few miles of trail that they followed could have been covered in less than two hours, but each track had to be positively identified, measured, photographed, and preserved. There could be no mistakes. I demanded that they treat this tiger case as if they were preserving evidence for a murder investigation, where someone's guilt or innocence rested on the proper reading of the tracks. The last thing I wanted was for our tracking evidence to be ripped apart in court or, worse yet, an innocent person deemed guilty.

The team, also accompanied by Tony, did great work. The gathering of the evidence was meticulous. Every track was marked, every hair sample collected, and every track measured. Ultimately, the tracks led to the tiger preserve compound. The evidence was very clear near the outer compound fence. Not only were there perfect tracks showing where the tiger had gotten under the fence, but there was a cluster of hair from the tiger on the fence where it had been pulled out. The hair could be analyzed for DNA and compared to that on the dead tiger. The team returned to our offices at the Tracker School around dusk, triumphant at what they had done. Considering that the tiger had crossed some difficult tracking areas, the team did well, but they were shocked when after the debriefing I told them that they had to return in the morning to finish the backtracking scenario.

Kevin was indignant when he heard that I was sending the team back. He could not understand why. He even argued with me, saying the hair on the fence also contained skin tissue, which would make the DNA test very accurate. Finally I asked the team why they thought I was sending them back but they hadn't a clue. In their minds they could not have done a better job. After all, there were in fact two teams of four people each and all of their findings agreed. Finally I told them why they had to return by posing a question. It was a question they should have asked themselves before they left the area, a question they would have asked themselves if they truly were not prejudiced. Yet in their minds they thought they were very open to what the tracks had told them.

I gathered the group together and asked them if they had cut track around the entire compound area. They looked puzzled at my question. Of course I knew that the answer was no. They seemed perplexed at my question until I explained further. I asked them if it was possible that a tiger from another location could have been lured to the compound by the scents of other tigers. It was obvious to them finally where the flaw in their backtracking analysis occurred. If they were truly not prejudiced in their analysis, then they would have thoroughly cut track around the entire compound, confirming that no other outside tiger had wandered to it and tried to get into the compound. I told them that this was the same question a defense attorney would ask us in court and could then use our answer to aid in breaking the case.

Without any argument or hesitation they returned to the compound the next day and finished the backtracking. There was no evidence that any other tiger had come to the area from another direction. It was clear that the dead tiger belonged to the owners of the preserve. We had preserved and recorded a clear line of evidence leading from the compound to where the tiger had died. This enraged the tracking team and the local community. The tiger compound was a horror story of filth, rats, poor fencing, and piles of excrement. The team was sickened by the conditions and literally hated the people responsible. They couldn't understand why I did not feel the same way they did. In fact, I told them that I felt sorry for the owners and they could not understand my feelings. After all, the poor fencing had caused the needless death of a tiger.

I told Kevin and the tracking team that I felt the same way they did but with one major exception. I have always taken pride in the fact that I could see the point of view from both sides and this case was no exception. Here was a group of people who were trying to save tigers. These folks took in tigers that no one else wanted and tried to give them a good home. They were poor folks and had to operate the preserve on a shoestring budget, largely from meager donations. They even fed the tigers by collecting road-killed deer. They would also go to many schools and give lectures on tigers and tiger preservation. Basically they were trying to do something good, but because of meager financing had not maintained the fence well enough. I also told the team that I would not want to see the preserve shut down. Rather I would want to see the work continued with greater financial help.

The more I spoke about the virtues of the preserve the more

they understood. Being a Tracker means also that we must be the same fence walkers in life that we are in following tracks. It would have been far too easy to condemn the preserve and hate those responsible. But by understanding the larger picture we can also understand the reasons behind the death. No, the owners of the preserve were not hateful people who were collecting tigers for personal gain, but sincere and loving people trying desperately to give abandoned and unwanted tigers a good home. Their only failing was not being able to afford to maintain the compound. We all agreed that we would pull together and help the preserve owners continue their work and help raise public awareness to their desperate need for donations and volunteers.

Still to this day we are all troubled by the death of that tiger. It was such a beautiful, healthy and strong animal. Its path to freedom only led to death. We could see in its tracks the truth. It was more like a lost kitten than a ferocious tiger. Realistically, as its hunger would inevitably grow stronger so would its threat to the community. I just wish that I could have been five minutes earlier on my initial arrival. I vowed from now on that I would add stalking to my next law-enforcement class. Knowing the art of stalking and invisibility would have given the police the ability to tranquilize the tiger. Stalking would also give my officers the ability to get close to fugitives and stray animals. Not that I expect to see another escaped tiger anytime soon, but as people's lust for exotic pets increase because of their greed, I no doubt will have to track these exotic pets again. I pray we are called in sooner.

TIME AND ETERNITY

THE CALL CAME in sometime during the middle of the night. There had been urgency in Joe's voice, which upset my father. Joe was a close family friend, a retired New Jersey State Trooper, who was responsible for getting me involved in my first tracking case many years earlier. In fact, Joe was so close to our family that we all called him Uncle Joe. To me, he was the epitome of a State Trooper. He was tall and very muscular with a bulldog face that was all business. Even though he was nearly seventy-five years old, it seemed like he had never fully retired. His attitude was that of an old drill instructor, yet Joe also had a softer side that very few people saw outside of the family. He had become an unofficial locator of the Tracker, as most law-enforcement agencies called me, and at times he hated that job. Basically, I was always so hard to find since I lived in the woods and wandered the country at random. Most times no one knew where I was or when, if ever, I would call home. This night Joe had been lucky because for the first time in nearly six months I was asleep in my own room.

Even when I was home, however, I rarely slept in the house. I hated houses, buildings, and rooms. Normally I would have been sleeping outside someplace, usually in a small shelter behind my folks' house. But I had been on the road for the past six months, most of the time in a full survival situation out in Oregon at the base of Mount Hood. And I had not come straight home from my adventure. For the past ten days I had been on a Vision Quest in the Pine Barrens. I had fasted for those ten days from food and the last two days I had purposely gone without sleep. At that point in my life I was still wrestling with the loneliness of having lost Grandfather, and within the same year Rick had died in a horseback-riding accident. I never felt so alone in my life. It seemed that no one spoke my language, felt my passion for wilderness, or understood why I would go alone into the wild places for such long periods of time. I had been hoping that this Vision Quest could in some way quench my loneliness.

I had only been asleep for a little over an hour when I was awakened by my father's voice. I could tell that he was aggravated in having to first find me and then wake me up. Apparently, assuming that I was not in my room, he had gone outside to the back shelter to find me. It wouldn't have been that bad but it was a bitterly cold January night, with high winds and a driving ice storm. It was aggravating enough that he had been awakened in the middle of the night, worse still that he found it difficult to wake me up. Yet in his own way he respected what I was able to do as a Tracker, though he would never tell me so to my face. After all, I had no job, no education beyond high school, and what seemed to him to be no ambition in life, at least not normal ambition. I had no idea how long he had been standing there trying to awaken me. In a way, not waking right up really pissed me off. I was always able to come fully awake, no matter how exhausted I was. Being able to wake quickly is necessary in any survival situation and I had broken that sacred rule.

My mind was still in a cloud, hearing only that Joe had called and that he would be picking me up in twenty minutes. I had to fight hard not to go back to sleep. I was puzzled because I began to think that it had been a dream, but when my mother came back into my room to see if I was getting ready, I knew that it was a horrible reality. Horrible because I was so exhausted and had no idea what Joe wanted or where I had to go. Questioning my mother, she just confirmed that he had called, had said it was very urgent, and that he would be picking me up. He had given no other details. All I knew for sure was that it must be a tracking case, but I had no idea of any other details. With tracking cases I could find myself going to anyplace in the world, in any topography and in any weather condition. I suspected that, given the urgency that my folks spoke of, someone's life was in danger. So too could it be some criminal that had just escaped. All I knew was that I had to be ready to go immediately.

My mother stayed with me as I got ready. She is a chronic worrier, especially so when I am headed out on an unknown tracking case. Yes, I know that she worried when I would go off into the wilderness with nothing, but after many years of this she was confident in my skills and abilities. The tracking cases, on the other hand, would really upset her, but she would never ask me not to go. She knew that many times I had pulled people out of the wilderness who would have surely perished, but this night she was frightened that I might be asked to track down some fugitive again. Those tracking cases really scared her in a big way. After all, at that point in my life I had already been stabbed twice, had received nearly a dozen concussions, and had lost several teeth in fights with fugitives. Needless to say she was very upset. But I think what upset her more was the fact that she knew I was extremely tired and the weather was horrible.

Another thing that bothered my mother was the way I would get ready and pack for any excursion. It made no difference if it was a tracking case or just a walkabout in the wilderness; my packing was nearly always the same. To the untrained eye, I packed virtually nothing, but for me it was bringing more than anyone could possibly need. I guess that it is a trait with all mothers, overly concerned that their children are properly dressed, especially facing the weather conditions that I was about to go into. Not knowing where on the face of the earth I would be going, I put on a pair of wool pants, a buckskin shirt, pocketed a small knife, and wrapped a blanket coat under my arm. It is rare that I would ever bring a knife, but I learned that on a tracking case it is best to have one so that time is not wasted making a knife out of stone.

What few people could understand about me is that the wilderness is my home. The earth provides everything I need and then some. After all, why pack anything if you are just going into your own house. For me, having a steel knife in my pocket is as good as camping in a recreational vehicle at a Disney World campground. Also, as a Tracker, I don't want to be bogged down by any equipment. A backpack, canteen, flashlight, or even a firearm when tracking fugitives are all liabilities. Equipment tends to slow progress and is so cumbersome when moving through thick brush. It's better to just go without it altogether. I did, in the end, agree to take a flashlight to appease my mother. She simply said that she felt I would need it so I took it and didn't argue. I've learned over the years that my mother's instincts are usually right.

Despite the late hour, Joe blasted his horn from the end of the driveway. I knew then that the urgency my father spoke of was very real. Joe, given that time of night, would have come to the door to get me and at least say hi to my folks. Before I could get to the door, he blasted his horn again, this time several times. Oddly, I found my dad waiting by the back door, fully dressed and ready to go. This baffled me because he never went on a tracking case with me. As far as I was concerned, he was disinterested in any of my tracking cases, at least until they were all over. For some reason, this night he wanted to ride with Joe and me. I had no idea if it was something he sensed about the brief phone call or a concern about the weather conditions I was going into.

In any case I didn't want to argue and let him come along. I know that once I hit the bush he would not follow me. It was common knowledge to everyone that I always tracked alone.

As we rushed to Joe's car a swirl of thoughts and anticipation filled my head. My greatest concern was why my father was going with me. As I said, I track alone. To me, anyone else accompanying me on a tracking case is an extreme liability. Whether it is the police or searchers, anyone with me tends to become a distraction. That's why my first order of business is to lose my backup, even when I am tracking the most violent fugitive. In most of the cases where I have had major fights with criminals it is because of the people who have been assigned to go with me. They tend to blunder about the wilderness, usually unable to keep up, and eventually alert the fugitive to our whereabouts. All I could think about on the way to Joe's car was how I was going to keep my father from going in with me, just in case he insisted. I was also worried that Joe might want to go along, though from his years of experience with me, I knew that would probably not become an issue.

As soon as we got in Joe's car he floored the accelerator. Before I could say a word, he began to tell me what was going on, which was not much. Apparently the people who had called him had little information and just told him to get me to the point last seen as soon as possible. We were on our way to a small airport and a private plane would be taking me up to New England. What Joe did know was that a nine-year-old boy was missing in horrible weather and in a large wilderness area. He also told me that the boy was diabetic and in a short period of time would slip into a coma if he were not found and given medication immediately. There had been massive search parties out all day, but not a single trace had been found. Even extensive air searches and dog tracking teams had turned up nothing. It became very obvious to me, given the horrid weather conditions Joe spoke of, that any search team would have trouble, even highly trained dogs.

It was during Joe's conversation that I realized my dad had called him back during the time I was getting ready to go. Apparently he had tried to talk Joe out of asking me to track. For some reason the weather was really bothering him. Yet, like my mom, he had confidence in my skills. What I think scared him was the fact that I had to fly to New England in a small airplane. He doubted very much if we were even going to take off, given the severity of the ice storm. I think the reason he went with me, at least to the small airport, was to make sure the pilot was not going to take any chances. It always uplifted my spirits whenever my dad showed any concern for me. Usually he would keep his feelings and concerns to himself, never showing any worry. It was always through my mother that I would find out how worried he had been. After all, he was a veteran of World War II and had seen and been involved in horrible sea battles. Nothing seemed to worry him, except for me.

In a way my folks are a reflection of the way most people think. To most, the wilderness is just that. A horrible place of struggle, especially when one goes in with nothing more than a knife. As I said, to me the wilderness is home, but even with my folks knowing my level of skill, there is still that fear of the unknown. I think that my father would rather face a major sea battle than to walk naked in the wilderness. But as Grandfather so often told me, "Survival of the fittest for the human animal is not so much how powerful one is, but what one knows." Another thing that I know bothered everyone, especially my folks, was the way I would withdraw within myself, apparently brooding, whenever I was going out on a tracking case. They just didn't understand that a Tracker lives part of the time in his mind and heart, and part of the time absorbed by the track. They take this as worry and brooding, not the introspection it is meant to be.

The ride to the small airport on this night was no different for me. I was wrapped totally in my thoughts, in a place somewhere between introspection and prayer. Yet it is not a time to second-guess the tracking case, or even digest the scant knowledge of the missing child that I did possess. The last thing that a Tracker wants to do is second-guess a tracking situation or to prejudice himself. A Tracker must always be a fence walker, never believing anything until the tracks prove it to be real. And because a Tracker is to never become emotionally involved in a tracking case, no matter how intense—to become emotionally involved is to distort clear thought and analysis—to many I always appear to be without emotion, made more of ice than any ordinary human. This night I was as cold and icy as the storm that raged outside.

We finally arrived at the airstrip. Fortunately the weather had cleared and stars began to sparkle through the black sky. Even with the break I felt apprehensive. I really hate aircraft in any weather, just as I despise any mechanized transportation. It isn't that I am frightened of crashing. My dislike of modern travel has more to do with the speed at which it moves. I always prefer to walk whenever possible. I get a close look at things most people have never seen. Planes are worse than cars because they are so far above the earth and so fast that everything is a blur. Time and distance are distorted and the earth remains untouched and misunderstood. Still at times like these, especially where a human life hangs in the delicate balance of time, I have no other choice but to get there as fast as I can.

It was not only me that wanted to get there fast, for everyone else wanted me there even faster. There is also a danger with speed and the Tracker mind. Tracking is not only a solitary pursuit, but also cannot be rushed. Rushing against time will only create mistakes, and mistakes can be more costly than prejudice. In the final analysis, rushing and then having to correct a time-related mistake only costs much more time. Part of the quietude I had this night was to keep from getting caught up in the rush. It is a type of meditation where one bends the time of man into the eternity of the earth. This time shift is difficult to achieve in this type of tracking case because both time and eternity were against the lost boy.

As we approached the airplane, which was already running and sitting on the runway, my mind raced back over something Grandfather had said about time. He once said, "Time is a myth, created by people to mark their place and position in life relative to each other, but the other dimension, a rhythm that goes on and on, without the need to mark place or position, that is eternity." I had no idea why this thought suddenly emerged from the

shadows of my mind. Yet I never question any sudden remem-
brance from Grandfather. Somehow, no matter how unrelated it
seems at the time to what I am doing, in the end it has great bear-
ing and insight. The only clue I had now to its importance was
the fact that this boy was running out of time, eternity, and hope.
I never equated the myth of time to myself at that time, though
now looking back I should have.

I walked to the plane in a cloud of thought. At first I didn't
hear Joe or my father tell me good-bye. In fact, it wasn't until
they said it a second or third time that I even grunted in
response. At this point my mind was withdrawn into the tracking case, the impending plane flight, and most of all Grandfather's statement of time. As I began to close the door, my father
ran to the plane and told me to get some sleep, any sleep. I know
that it must have seemed to him that I was about to go into battle, and from his war experience he had learned to grab sleep, no
matter how short, anytime he could. I don't know where that
statement came from, given all the things he could have said, but
I've learned, as I did with my mother's advice, that any advice
coming from my dad, especially at the spur of any trying
moment, should be taken seriously. In response I just told him
that I would try and closed the door.

My pilot was a weird sort, far too awake for this time of night.
He had a gift of gab, as my Scottish grandmother would say, and
subsequently didn't shut up until we were well in the air. Judging
from the takeoff his experience as a pilot was also questionable;
such was the cavalier way he handled the airplane. Fortunately,

as we flew into the night and the conversation jumped about I learned that he was a Vietnam pilot and any apprehension I had vanished. I should have just turned off the headphones and told him I needed to sleep, but his stories of the war were enthralling and I wanted to hear them. Even though I was never a soldier, I had spent some time in Nam, Cambodia, and Laos, supposedly teaching tracking to various elite troops. To my amazement he knew of me and the men I had trained. It was great to share some stories of covert operations with someone who really knew.

I must have dozed off for the last fifteen minutes of the flight. All I remember is seeing the runway in the distance and the next thing I knew I was on the ground with the door open. Again, it took a while to wake me up. I was startled by the fact that the one waking me was not a member of the police department or sheriff's office, but a member of the boy's family. This was very upsetting because I instinctively knew that it was not the police that called me in but the family of the child. This is always so difficult because unless the law-enforcement agency knows of me and calls me in, I am usually met with animosity. Typically, back then, though I had worked with countless law-enforcement agencies, the FBI, and other organizations, not everyone had heard of me. Not only did I then have to prove myself to the police, but also I might not even be allowed to go in to search for the boy at all.

I couldn't really blame any police department for their lack of cooperation, especially when they didn't know who this Tracker is. After all, tracking to the untrained observer is an ancient practice, more art than science. So too can I imagine how they feel when someone from another state, and a civilian at that, comes onto their turf and does something that they cannot do themselves. The factors of the ticking clock, the horrible killing weather, and the boy's medical condition were now compounded by the fact that I might be facing opposition from an entire police force. I realized then that I had my work cut out for me. I was also pissed off at Joe for not telling me. Yet I know his motivation. If he had told me I might not have gone at all unless I had the permission of the authorities. Time would not allow for this option.

I resolved myself to the fact that I was there and all I could do was the best I could no matter what the circumstances. After all, a child's life hung in the balance, not politics, and that became my main driving force. I was hoping too that the pressure of this search-and-rescue operation would have worn down the authorities. Once they have exhausted all of their resources and have public pressure they tend to welcome outside help of any sort. Even psychics tend to be welcome when the trail, if any, has gone cold. I would not allow myself the luxury of worry or concern over this potential problem. All I could do was hope for the best and find out as much information as possible from this family member. This proved to be a lesson in futility because the family member was the husband of a distant cousin of the child and knew little more than Joe.

My first order of business was to interview the child's mother and father. It is so important to a Tracker to learn as much about a missing person as possible, even before the actual tracking begins. As far as I am concerned, the actual tracking begins when you question all of those people close to the missing person. It was not just the physical traits and characteristics that I was interested in but also the child's interests, his mental and emotional quirks, and any other bit of information the family could give me. What few people understand is that a Tracker must become the animal or human they are tracking. By getting into the boy's mind, I could better track him right from the beginning. I also know that this can be a very painful process for the parents, especially when a child is missing. Their anguish, suffering, and pain tend to distort what is true. That's why I always go outside of the family, to friends, teachers, and even mild acquaintances to confirm information.

All too often I will find that when the parents of a missing child say that he or she is well adjusted and happy, his friends paint a far different picture. I've heard so often of a child's happiness, according to the parents, only to find that his friends tell of suicidal tendencies and depression. Families tend to be very good at setting aside the truth and creating utopian images. As I've said, a Tracker is a fence walker and does not believe anything until the tracks reveal the truth. It baffles the authorities when they accompany me on these fact-finding missions because, to me, the person's mental and emotional state can be more important than the very shoes he is wearing. It always surprises me how little parents know of their own children or even know of each other. I've found in life that achieving awareness of the environment or even their own families are not top priorities for modern man. In fact, it tends to be nonexistent and very prejudicial.

I could tell that my driver was not thrilled with taking me first to the boy's house to question his parents. Neither were the parents for that matter. As far as they were concerned, I should have been taken to the point last seen. Even though I never charge any fee for a tracking case or accept even a meal, because they send a private plane they feel like they somehow own me. Worse still, they try to tell me what I should be doing or where they think the missing person might be. They also complain about the police, generally stating that they do not know what they are doing. All I can do at this point is to nod my head and listen. Somehow I always become a whipping boy and a place to vent anger over the search-and-rescue operations. Of course, this is the first thing that I had to sit through before even asking any of the pressing questions that I needed answers to.

As I expected, the parents of the boy were not asleep, though both were heavily medicated. No one can imagine the pain and anguish parents go through at the loss of a child, and these two were no exception; both were basket cases. After the usual formalities, I set about asking my questions. I typically ask a question and then look away at the house or surroundings as I listen intently to the answer. This tends to be very annoying to people because they assume that I am disinterested, but that is far from the case. I certainly do watch body language and pay attention to word choices, but I also observe very carefully the house or surroundings. I learn much by seeing how people live, the types of things they have in their homes, the pictures that are around, and little nuances that would escape most people. I knew almost immediately that this couple had been divorced and that it was a rough divorce at that.

When I asked them how long they had been divorced, they looked shocked. After all, no one had told me anything about them and it wasn't common knowledge. To me it was blaringly obvious, from the pictures on the wall to the lack of men's clothing around the house, or at least men's clothing that would fit the father, and many other things that spoke of a horrible divorce. I also knew that this family was once very well off financially but were now, probably due to the divorce, living on a very tight budget. I could tell from their reactions that they did not like me reading between the lines of their lives. There were things they just didn't want people to know about their private lives and I seemed to strike a nerve with my deeper questions.

Armed now with an intimate knowledge of the boy and the family in general, I was rushed to the point last seen, which was not far from the house. As usual there was an assembly of vehicles, dogs, horses, a small communications van, and several police cars, both local and regional. Even though it was very early morning the compound was full of activity. Search parties were going out in all directions, orders were given, and late-night searchers were returning exhausted from the hills. To someone who had never been involved in a search before it would seem like utter chaos, but this level of frenzied activity is normal. In the center of all this sound and motion I immediately spotted the people running the operation. It was clear that they had had little sleep and were at the edge of physical and mental exhaustion.

The boy's father took me right up to the lieutenant and

introduced me as the Tracker. At this point of introduction my heart was in my throat, especially because of the look on his face. He smiled at me coldly and asked, "What's a Tracker?" and just as I was about to explain I felt a hand slap me on the back. I turned to find a familiar face. The man behind me was the sergeant on duty and a man I had worked with twice before on criminal cases. Seeing him was like taking a breath of fresh air. I could feel the relief wash over me as he explained to the lieutenant that I had helped the department read forensic photographs of a murder case several years earlier. It was then too that the lieutenant recognized me, or rather my work, and shook my hand.

It was obvious that they were both relieved. I know they must have put up with a lot of criticism from the boy's folks previously, and when I was brought in by the father it created an apprehension. Now they knew that I was on their side and that I could somehow help them. This was such a relief to me to have the backing of the law-enforcement officials. Without that it would only make tracking more difficult. Now at least I could go in with the blessing of the department. In the past, some of the greatest obstacles I've had to face were reluctant and skeptical police departments. At least now I did not have to prove my abilities and myself to them. They were already won over.

I could feel my internal clock ticking loudly as the sergeant led me to the point last seen. As we walked we did not reminisce about old times but concentrated instead on the remaining facts of the case from the police point of view. To them, the boy had run away in an attempt to bring his parents back together. He had gone into the woods for what was to be only a short but frightening time, but in the process had gotten lost. He had been gone now for a full day and night. The weather had been horrible and deadly cold. Worse still, he would by now be in need of medication for his diabetes. There was still a faint hope in the sergeant's voice, but I could tell that it was very faint.

We arrived at the last place one of his friends had seen him the day before. As usual it was tracked over by countless searchers, dogs, ATVs, and horses. I knew that it would be fruitless to begin here, for even if I found a partial track the next would be obliterated. I had to do what I always had done: get the hell out of the heavily searched areas and into the deeper wilderness. I had to find the perimeter of the heavily searched areas where the tracks, if any, would not have been tracked over. Once outside this area I could cut track, a process of cutting perpendicular to the search area and picking up any tracks leading out. This process could take days if done blindly. What I did have on my side was an intimate knowledge of the lost boy.

My sergeant friend knew enough about me to let me track alone, but breaking my own rules I asked him to come along. I told him that as soon as I found the trail he should go back, however, and alert the searchers to the direction I traveled. The reason I asked him along was simple: If or when I did find the tracks I would have sufficiently cut down the entire search area to the expected line of travel. This would cut the search area down to only a few miles instead of the hundred square miles they were now searching. As we set off I warned him not to allow any searchers into my area for a full twelve hours. I did not need to have the tracks destroyed by well-meaning search-and-rescue groups.

It was midmorning when we set off, or so I assumed. The sky had grown very dark, the winds began to howl again, and temperatures plummeted. I picked my way along the trial, watching the searchers' tracks slowly diminish in number. Finally, after two hours of travel and intense scrutiny, I finally found the track of the boy. Though it was only a partial heel track I knew it was his. It was the right size, the right time, and the right pressure releases. I knew the boy's pressure releases by feeling inside one of his old and worn shoes back at his home. I was certain this partial track was his and I was headed in the right direction.

As soon as I spotted the track I paused, leaned down, and touched it for a long moment. Often the hands are more sensitive than the eyes. I can feel the track, and combined with what I am seeing I can add another dimension to the print. I could see and feel his exhaustion, his confusion, and his fears. He must have realized long before this point that he was lost, for a sense of panic was beginning to set in. It was at this point, as with all tracks of animal or human, that I become the being I am tracking. Yet there is a fine line that a Tracker cannot cross. We must remain aloof, almost cold and unemotional, for if we give in too much to the track we can feel the pain, the exhaustion, the confusion, and the fear. It is a constant struggle within the deeper recesses of self.

At this point the sergeant and I parted ways. I could tell part of him wanted to stay with me, but the other part, the more deeply fatigued searcher he had been, wanted rest. It was in my hands now. I thought, as I walked away, that he would have made great company if he had been more rested. I know that he would have loved to find the boy with me, but we both knew that was out of the question. I could not break my own rules. His company would have been great, but it would eventually have become a distraction and slowed the inevitable process of tracking. We walked away from each other without a word of goodbye or even a glance backward, at least on my part.

I reached the ridge sometime around noon. Here on the more exposed areas the wind and cold were very intense. No tracks, except for those of animals, continued along the path I was following. It was time for me to cut track. Something inside me pulled to the right. That something was the voice of Inner Vision, a sense of knowing the boy, of becoming the boy, and what the landscape was telling me. I had to pay attention to that voice, for to disobey and follow the logical would be a disaster. The logical approach would have been to go to the left. That is the way the trail went and it also led downhill. To the right the terrain was steep and tangled in brush. If the boy were thinking clearly he would have not gone that way. He was not thinking clearly. He was in a place of panic.

I cut track for the better part of the early afternoon, ready to give up my line of search and head back to the more logical approach. It was at this point of indecision that something caught my eye, a glint, far in the distance and of unknown origin. I pushed through the heavy brush, away from my original track cut line and toward the glint. It was tough to follow, for the sky was deeply overcast and the glint very faint. I had to keep my eyes on the overall landmarks and trees to get to the exact area, for the glint was also erratic due to my ever-changing position. Something deep inside drove me on. Even though my mind was logically thinking it was the glint of garbage, I felt that it was more.

I found the boy's tracks before I found his small backpack. I picked up the backpack, knowing now that the child was in a state of panic that was fast becoming shock. A rational person would not have discarded his backpack, but at this point I've witnessed people discarding full canteens in the desert, walking into trees, or even stepping off cliffs. The boy was on the edge, driven by panic, shock, cold, and the oncoming coma of diabetes. I could see it in his tracks and feel it in his discarded backpack. A sinking feeling came over me when I searched his pack and found no syringe. His folks had told me that he always carried one, but there was no evidence around that one had been used. I prayed that he had put it in his pocket for later use, but that was doubtful.

Tracks to me are not only a window to the past, but a window to one's very soul. I could see in the boy's tracks the frustration he felt when he rifled though his pack, unable to find the syringe. I could see the exasperation and anger in his tracks, an anger that finally gave way to a blind panic. I followed his tracks easily now as they headed down the hill. He blundered about, tripped frequently, broke through low brush, and intermittently broke into a panicked run, only to fall again. In various places I found his blood on small twigs, for his face had been badly cut in an earlier fall. I also found bits of his hair and fibers of his coat. The fibers were not of a sturdy wool outdoor coat, but a light nylon jacket. It was obvious that he was underdressed and not wearing the coat his folks had told me he should be wearing. Given the conditions, these details held lethal implications.

The oppressive weight of time now overwhelmed me. Deep fatigue from lack of sleep was setting in and I had to struggle to think clearly. My reactions had slowed along with my thoughts and I teetered on the edge of becoming emotionally caught up in the boy's fear. I had to push myself hard to make headway. I had arrived at the point of "dead tracking." Dead tracking is the act of following someone faster than he is moving. Dead tracking presses the limits of a Tracker, for if he fails the trail ends in death. I knew from the age of these tracks that I was eight hours behind the boy and I had to cover those eight hours as fast as I could. Given all of the horrid conditions, the child would not last much longer. If he made it through the oncoming night he might have a chance. I knew by the nuances of nature and sky that the weather would change by morning and become warmer.

I tracked into full darkness. The boy's trail was now chaotic and showing signs of profound shock. There was no reason behind the choices he made, of which trail to follow, or even to avoid a frozen shallow pond. His soaking-wet tracks emerging from the other side showed signs of hypothermia setting in, along with the onset of slipping into a diabetic coma. I could no longer see the tracks but had to feel them with my hands. I used the flashlight my mother had given me but sparingly. I knew that it would burn out through the night if I left it on. I used it only to confirm obscure tracks and resorted to feeling for the rest with my hands. Long ago Grandfather had taught me to track with my hands nearly as fast as I could with my eyes. Now again this technique was paying off.

Eventually the flashlight burned out and the trail became indistinct to my touch. I fumbled with making a torch, but my fatigue had overwhelmed me. I had to resort to lying on my belly and following the tracks with the gentlest touch I could muster. My hands were frozen and numb, further intensifying my own pain and obscuring the touch. I dragged on for the better part of a mile, feeling the age of the boy's tracks now only an hour ahead of me. As I tracked I had to fight sleep. I had to stay awake and alert to every movement of the forest. I could not afford to make a mistake, not when I was this close.

I awoke to the faint chirping of birds, and I was enraged. I had fallen asleep. Time I could not afford. I had given in, given up, and I was angry at my frailty. I could easily have dismissed the anger—after all, I had been up for nearly four days at this point—but I could not tolerate that kind of frailty, especially when another human life hung in the balance. I exploded to my feet like a seething beast and began to run along the boy's tracks. What had been so difficult to feel during the night was now so clear and easy. My mind began to punish itself. Even though I had fallen asleep for less than an hour it was far too much time wasted, more time than I could bear or the child could afford.

As I ran along his tracks I could see that he was dying, I could feel him dying, just hours before. He blundered and stumbled.

He had no thought, just action and reaction. I was no longer following single tracks but whole sections of his trail. I could feel the concentric rings of nature coming from the distance, telling me that he was there by a distant rock outcropping. There was no doubt in my mind where he was located; the birds just don't lie. I drew up every ounce of strength I had left and sprinted to the outcropping, hoping for the best but knowing that I was too late, possibly one sleep too late. Even before I got to him I hated myself, I hated time, and I loathed eternity.

I found him huddled under the lip of the rock outcropping, looking very much like he was sleeping. I rushed to his side and pulled him close. His body was still warm but limp. He had no pulse. His face was ashen gray and his eyes partially open. I could see the fear in them and also a hint of faint peace, as if somehow he had been released. I tried a frantic effort to revive him, blundering more than doing any good. I tried to warm him, I tried CPR, I tried everything I knew. I just could not accept the fact that he was dead. I cursed myself and my utter failure as a Tracker. I was caught up in my own madness. I could not think or accept his death.

I was so enraged. I remember vaguely cursing, pounding the ground with my hands, crying, and calling up to the Creator. I could not think clearly or make sense out of anything. I couldn't even decide whether to carry the boy back to base camp or leave him there to preserve the evidence. Had a crime been committed by the boy, or upon the boy, or was the only crime that I had given in to sleep? In a very feeble way, I marked off the area with sticks, knowing full well that my attempts to revive the boy had destroyed any evidence of his death. I should have left him there and gone back for help, but I could think of nothing more than to punish myself.

It would have been easier to walk out alone, given my state of exhaustion. I could then direct the police to the body and be done with the tracking case. But that would have been the easy way out. I carried the boy over my shoulders. I wanted to feel his lifeless body reminding me constantly of what a loser I was. I would not allow myself the luxury of rest, for all I wanted was to punish myself, hoping that I could also die rather than face the child's parents. I could not deal with another child's death or the pain and suffering of those he left behind.

My mind would not let go. I began to criticize my every move and action. I cursed every rest I took while tracking. Every pause or hesitation was a criminal act as far as I was concerned. I had let down the boy, his parents, myself, and most of all, Grandfather. He would have been ashamed of me. I know in my heart that Grandfather would never have quit, paused, or fallen asleep. He had blessed me with his teaching, his skills, and now I had let him down. I had never felt so unworthy in my life. I had disgraced the honor of the ancient Trackers. Humbled, humiliated, and enraged I stepped to the bushes just outside base camp and lay the boy down. I prayed for him as best I could in the blind coffin of my anger.

I glanced over at the sergeant and he knew instantly that I had found the boy. All I remember is pointing to the body and then nothing. I awoke in a clinic many hours later. Standing was difficult, but orienting myself was more difficult still. I walked into the outer office and to my surprise the sergeant was waiting. I could tell that he had gotten no sleep but had opted to stay by my side until I awoke. He looked at me knowingly and said, "The doctor said the boy was dead at least four hours before you found him." I told him that I could not accept that. I had wasted too much time and had failed Grandfather.

We drove to the airport in silence. The only thing I asked was that they not mention my name to the media. I never accept any credit for a tracking case, but give the credit to the police and search teams instead. The most I would ever say is, "Tell them that the Tracker helped." All I could think was that the Tracker killed the child because of his frailty. The rest was a blur of action and reaction, of silence, the drone of plane engines, and not a word from Uncle Joe. I know that the lieutenant had briefed him long before I had arrived back at the airport. As I walked away from the car, Joe said, "Don't beat yourself up. He was dead long before you got to him. If it weren't for you he might never have been found."

In the days that followed the tracking case, something shifted in me, something changed. To this day I still punish myself whenever I think about that case. Even though logically I know that it was not my fault, I still could have pushed every ounce of time I had. I vowed never to quit, never to give in to sleep, never to slow or pause in my quest to save a life. I vowed to Grandfather and the Creator that I would die tracking rather than fail. Time and eternity will forever remain fused in my mind and heart as one. For a Tracker of life is always in a race between the two dimensions of time and eternity and must sacrifice all so that someone's time does not run out to eternity.

At the age of eight, Tom Brown, Jr., began to learn tracking and hunting from Stalking Wolf, a displaced Apache Indian. Today Brown is an experienced woodsman whose extraordinary skill has saved many lives, including his own. He manages and teaches one of the largest wilderness and survival schools in the U.S. and has instructed many law enforcement agencies and rescue teams.

As you know from reading this book, sharing the wilderness with Tom Brown, Jr., is a unique experience. His books and his world-famous survival school have brought a new vision to thousands. If you would like to go further and discover more, please write for more information to:

THE TRACKER

Tom Brown, Tracker, Inc.
P.O. Box 173
Asbury, N.J. 08802-0173
(908) 479-4681

**Tracking, Nature,
Wilderness Survival School**